The Fungal Web

The Underground Networks That Coordinate Forests

Team Trex

Published by NobleTrex Press

For permissions and other inquiries, write to:
P.O. Box 3132, Framingham, MA 01701,
USA

Contents

Introduction

Stand in the middle of a forest. Close your eyes. What do you perceive? You might hear the rustle of a canopy agitated by the wind, the snap of a twig, or the trill of a thrush. You smell pine resin and damp earth. But mostly, you perceive stillness. To the human eye, a forest is a collection of stoic, solitary individuals-trees standing in rows, competing for light, rising in vertical silence. We see them as architectural columns, separate and distinct, their branches occasionally touching but their lives essentially parallel.

This view is a majestic illusion.

If you were to dissolve the soil beneath your feet to reveal the biological machinery hidden within, the stillness would vanish. You would be standing not on a solid floor, but on a chaotic, throbbing ocean of connectivity. In a single teaspoon of healthy forest soil, there can be several miles of fungal filaments, known as hyphae. These are not merely rotting agents or pathogens; they are the cables of a subterranean internet, a physical network that links the roots of one tree to another, and another, until the forest floor reveals itself to be less like a collection of statues and more like a single, self-regulating superorganism.

This is the fungal web. It is the oldest social network on Earth, a sprawling, decentralized infrastructure that has been coordinating the biosphere for over 400 million years. It is the reason forests exist as we know them.

For centuries, biology was dominated by a narrative of ruthless competition. The "survival of the fittest" was interpreted as a gladiatorial contest where every organism fought for its own slice of the resource pie. In this view, a tree was a fortress, hoarding its sugars and water, guarding its territory against neighbors. But that dogma began to fracture in the late 19th century and shattered completely in the last few decades. We now know that beneath the competitive canopy lies a cooperative basement.

The realization that plants communicate and share resources underground is perhaps the most significant shift in our understanding of ecology since we mapped the food web. It challenges our definitions of individuality. If a Douglas fir pumps carbon into the roots of a nearby paper birch, and that birch later reciprocates when the fir is shaded, where does one tree end and the other begin? If a dying sapling offloads its remaining resources into the network before it withers, is it an act of altruism, or is the network simply reclaiming its assets?

To understand this hidden universe, we must look at the architects of the system: the fungi. When we think of fungi, we usually picture mushrooms-the fleeting, fleshy fruit bodies that pop up after a rain. But the mushroom is merely the reproductive organ, the apple on the tree. The tree itself is the mycelium, a vast, white, cobweb-like entity living inside the soil, inside rotting logs, and, crucially, inside the roots of plants.

This partnership is called *mycorrhiza*, a term coined by German botanist Albert Bernhard Frank in 1885, combining the Greek words for "fungus" and "root." Frank was ridiculed by his contemporaries for suggesting that trees might actually *need* these fungal infections to survive. Today, we know he was right. Roughly 90 percent of all land plants form these relationships. The deal is a biological barter system of staggering complexity: the plant, a master of photosynthesis, captures energy from the sun and converts it into sugars. It sends a significant portion of this "liquid carbon"-sometimes as much as 30 percent-down into its roots to feed the fungus. In return, the fungus, a master of mining, extends its microscopic hyphae into the soil pores that thick tree roots cannot reach, dissolving rock and scavenging for phosphorus, nitrogen, and water to pump back to the plant.

But the story does not end with a simple one-to-one trade. Because fungal hyphae are promiscuous, connecting to multiple trees of different species, and trees are equally gregarious, hosting dozens of different fungi, a physical network emerges. This is the architecture of the "wood-wide web," a term popularized by the journal *Nature* in the late 1990s to describe the findings of researchers who proved that carbon could move between trees.

The implications of this connectivity permeate every aspect of our lives, even if we never step foot in a woodland. These networks act as the stabilizing force of the global carbon cycle. They determine how forests react to climate change, how they recover from fire, and how they scrub the atmosphere of the greenhouse gases we emit. When we clear-cut a forest, we are not just removing timber; we are tearing up a biological circuit board, severing the links that keep the soil from turning into dust. Understanding these networks is no longer just an academic curiosity-it is a prerequisite for planetary stewardship.

As we peel back the layers of the forest floor, we find that the "wood-wide web" metaphor, while useful, is somewhat incomplete. It suggests a sterile, digital highway of information. The reality is far more visceral. This is not just a fiber-optic cable; it is a bustling bazaar, a zone of negotiation, and occasionally a battlefield. The exchange of nutrients is not a charity; it is a market economy where prices fluctuate based on supply and demand.

We will explore the mechanism of this trade, observing how fungi can hoard resources to drive up the "price" of phosphorus, or how trees can cut off "cheater" fungi that take sugar without providing nutrients in return. We will look at the work of Toby Kiers and others who have demonstrated that these organisms make decisions-economic decisions-without a brain. It turns out that the wisdom of the forest is not held in a central processing unit, but is emergent, arising from millions of local interactions between root tips and fungal threads.

This shift in perspective requires us to adopt the tools of network science, the same mathematics used to map the internet, social media connections, and air traffic control. By viewing the forest as a graph of nodes and links, we can identify the "hubs"-the massive, ancient trees that Suzanne Simard and her colleagues have identified as "Mother Trees." These elders are the gravitational centers of the forest, linked to hundreds of younger trees. They act as reservoirs of genetic diversity and resources. When we apply this lens, we see why selectively logging the biggest trees can cause a disproportionate collapse in the forest's biodiversity. It is akin to removing the major airports from a flight network; the whole system loses its redundancy and resilience.

However, science is a practice of rigorous skepticism, and the idea of the "helpful forest" has occasionally outpaced the data. We must be careful not to anthropomorphize trees as benevolent beings holding hands. We will interrogate the evidence for the more controversial claims: Do trees really "talk" to warn each other of beetle attacks? Is the transfer of carbon between trees a deliberate act of aid, or is it a passive flow down a chemical gradient, driven by the physics of source and sink? We will examine the innovative methods scientists use to answer these questions, from stable isotope tracers that light up the movement of atoms like distinct dyes in a river, to the genetic barcodes that allow us to identify who is connected to whom in the black box of the soil.

The deeper we go, the more we realize that the fungal web provides a blueprint for resilience. In a world defined by rapid climate instability, the redundancy of these underground networks offers a buffer against shock. A diverse community of fungi can help a forest withstand drought, repel pests, and hold the soil together against erosion. But this resilience has a breaking point. We will examine how industrial agriculture, nitrogen pollution, and modern forestry practices have disrupted these ancient alliances, creating "silent" soils where the conversation has died out.

Yet, this is not a eulogy for the woods. It is a manifesto for a new way of working with them. By understanding the underground network, we are discovering new methods for restoration. We are learning that planting a tree without its fungal partner is like putting a fish in a tank without water-it might survive for a while, but it will not thrive. We are seeing the rise of "assisted migration," where scientists match trees with specific fungal guilds to help them survive in migrating climate zones. The future of forestry lies not in managing the timber above, but in stewarding the web below.

The fungal web invites us to see the world differently. It asks us to look at a tree and see not a wooden pillar, but the tip of an iceberg, the visible protrusion of a vast, subterranean life support system. It challenges the Western philosophical tradition that prioritizes the individual over the collective. In the forest, the individual is a myth. Survival is a collaborative act.

The ground beneath us is alive with signals, trade, and memory. It is a world of ancient alliances and microscopic miners, of sugar rivers and chemical whispers. To understand the forest, we must first learn to see the invisible threads that hold it together. We must descend into the soil.

Chapter 1

The Pulse Beneath the Floor

Beneath the apparent stillness of a forest walk lies a chaotic, biological superhighway where survival depends on negotiation rather than isolation. This chapter peels back the earth to reveal how centuries of viewing trees as solitary competitors obscured the vast, pulsing vascular system that physically connects the forest into a coherent whole.

1.1 The Myth of the Solitary Tree

For most of the modern history of forestry, a tree was treated as a kind of upright landowner. It held its plot of soil and air, defended its crown against encroaching neighbours, and paid for every gram of wood with private earnings from its own leaves. If it grew quickly, it was a winner; if it grew slowly, it had been outcompeted. The forest, in this view, was a crowd of individuals forced into proximity, and the scientific task was to measure their rivalry: how much light one tree stole from another, how far roots spread to intercept water, how density throttled growth.

It was a useful model. It made forests legible to management. A stand could be thinned like a garden bed to reduce competition; a species could be favoured or suppressed by controlling who gained access to sunlight. Most early silviculture—the practice of tending forests—grew up alongside the mathematics of yield and the economics of timber. In that world, *interaction* meant interference: one tree casting shade, one root system drawing down moisture. Anything that looked like sharing was either a mistake, a curiosity, or a sentimental projection.

Yet the forest itself kept producing awkward facts. Some were visible to anyone who has knelt beside an old stump. The stump, cut flush to the ground, sometimes remains alive for years. Its cambium—the thin living layer that makes new wood—stays moist. New tissue forms over the cut surface. In a few cases the stump even adds growth rings, as if still counting seasons. It cannot photosynthesize. If it persists, it must be subsidized. Foresters could explain this away as stored reserves, but the longer-lived stumps made the explanation creak. Another awkward fact lay in the way seedlings behaved beneath a closed canopy. A young conifer can linger in the shade for a decade or more, a dark green patience, barely adding height. Then, after a gap opens in the canopy, it surges. The solitary-tree model interprets this as simple release from competition. But the release is often too coordinated, too swift, too dependent on the chemical and microbial state of the soil, as if the seedling has been kept alive by more than its own meagre photosynthesis.

Even the language of old forestry hinted at something beneath the competitive story. Stands were called *even-aged* or *uneven-aged*, as though the forest were a demographic structure. Trees were said to be *suppressed* or *dominant*, as though status mattered. But the textbook mechanisms remained aboveground: light and space. The soil, by contrast, was treated as scaffolding. Roots were pipes. Nutrients were chemicals dissolved in water, to be absorbed wherever encountered. Fungi, if mentioned at all, appeared as pathogens, decay agents, or culinary curiosities.

The first cracks in this picture did not come from a grand new philosophy of cooperation. They came from method. Once a scientist can mark a molecule and watch where it goes, old assumptions become testable in a way that arguments alone never achieve. Radioactive tracers and stable isotopes are, in effect, dyes for the invisible economy of life. They let you tag carbon in one plant and ask, without metaphor, whether any of it turns up in another.

A concrete early clue came from experiments that were as much engineering as ecology. Imagine enclosing a branch, or a whole seedling, in a transparent chamber and flooding it with carbon dioxide that carries a distinctive signature—a radioactive isotope such as ^{14}C, or a stable isotope such as ^{13}C. The plant photosynthesizes as usual, turning CO_2 into sugars, but now those sugars are labeled. If the labeled carbon later appears in roots, in surrounding soil, or in a neighbouring plant, the movement is no longer a story: it is a measurement. Early tracer work in plant physiology had shown that carbon moves rapidly within a single plant, from leaves to roots and back again. Extending the method to multiple plants, in soil, was harder. Soil is a chemical fog filled with microbes that also respire and trade carbon. Roots can touch. Water can carry dissolved compounds. To claim that one plant has donated carbon to another, you must rule out a parade of alternative routes.

A field anecdote helps explain why these experiments mattered so much. In the interior of British Columbia, foresters and ecologists faced a practical problem: how to regenerate conifer forests after logging without losing productivity. Douglas-fir and paper birch often grow together. Birch was sometimes treated as a weed tree, competing with conifers for light and water. The management impulse was straightforward: remove the competitor and free the crop species. But some foresters noticed that conifer seedlings did not always thrive when birch was eliminated. In some sites, mixed stands seemed healthier, more resilient, less prone to failure. The orthodox explanation was microclimate: birch could shade seedlings from heat, or improve soil with leaf litter. Those are real mechanisms. But beneath them lay a more provocative possibility: that birch and fir might be linked belowground in ways that changed the very meaning of competition.

Suzanne Simard and colleagues put this possibility into the ground and insisted on an answer. In the late 1990s, they used reciprocal isotope labeling in the field to test whether carbon could move between paper birch (*Betula papyrifera*) and Douglas-fir (*Pseudotsuga menziesii*) when the roots were associated with ectomycorrhizal fungi—fungi that form a sheath around fine roots and extend outward as hyphae. The key result, reported in 1997, was not simply that labeled carbon could be detected across species; it was that transfer could be bidirectional and sensitive to conditions like shading, consistent with a source—sink logic rather than a sentimental one. A plant with abundant photosynthesis behaves as a source; a shaded plant behaves as a sink. When the researchers manipulated light, the direction and magnitude of net transfer shifted. The forest floor, it seemed, was not only a battlefield. It was also a market with fluctuating exchange rates.

For readers accustomed to thinking of trees as self-contained organisms, the idea of carbon moving between them can feel like a category error. Carbon, after all, is the stuff of a tree's body: wood, bark, roots. Why would a tree pay any of it out? The first answer is that the currency is not wood but sugar—photosynthate, soluble carbohydrates that flow through the phloem like sap and can be diverted, leaked, or traded. Plants routinely export sugars to their roots to feed symbiotic partners. Mycorrhizal fungi, in particular, are not charitable residents: they are paid. The plant supplies carbon; the fungus supplies nutrients and sometimes water. That exchange is ancient and widespread enough that many plants struggle to thrive without it. Once carbon is being exported into a fungal partner, the possibility arises that the fungus might deliver some of it elsewhere, especially if the fungus itself connects multiple plants.

The second answer is that not all carbon leaving a plant is a deliberate gift. Roots exude carbon compounds into the soil continuously, a kind of chemical halo that feeds bacteria and fungi. Cells slough off. Tiny wounds leak. Microbes metabolize these compounds and respire them as CO_2, which other roots can take up. Carbon can also move as dissolved organic compounds or as CO_2 in soil water, and can be refixed by photosynthesis after it diffuses into the air. If an experiment detects labeled carbon in a neighbouring plant, skeptics can reasonably ask: *through which route, exactly?* This is why the technical debates around common mycorrhizal networks can be both intense and healthy. Nature is full of pathways. Proving the fungal route requires clever designs: physical barriers that allow hyphae but not roots, isotopic signatures paired with fungal identification, time series that separate fast diffusion from slower biological transport.

This methodological caution has become more prominent as the idea of the *wood-wide web* has escaped the laboratory and entered popular culture. The phrase is catchy, and it flatters a human hunger for community: a forest as a collective intelligence, older trees nurturing their young, a society beneath the leaf litter. The trouble is that catchiness can outrun evidence. Justine Karst and Melanie D. Jones argued in 2023 that common claims about common mycorrhizal networks have been amplified beyond what the peer-reviewed literature supports, in part through positive citation bias: later papers repeating attractive interpretations without the same level of qualification. David G. Robinson and Christian Ammer made a related point in 2024, warning that personifying trees and fungi—calling them altruistic, maternal, or purposeful—can turn metaphor into mistaken fact. These critiques do not require rejecting underground connectivity. They ask for a more disciplined imagination: one that can be fascinated without being gullible.

How, then, to challenge the solitary-tree myth without replacing it with a fairy tale?

A good starting point is to separate three ideas that are often blended. First, *mycorrhizal symbiosis*: many plants partner with fungi, trading carbon for nutrients. This is solid ground. Second, *physical connectivity*: a single fungal individual—a genet—can colonize multiple plants, creating a network of hyphae that physically links root systems. This, too, has strong support in several ecosystems, especially those dominated by ectomycorrhizal fungi. Third, *functional consequences*: resources, signals, or water move along these links in ways that measurably affect plant fitness, community composition, or forest resilience. This is where the evidence is most variable, and where the most alluring stories often live.

The solitary-tree dogma falters as soon as the first and second ideas are taken seriously. A tree does not end at its bark. It extends into the soil as a set of fine roots colonized by fungal tissue. The actual absorbing surface of a tree, in many cases, is not root epidermis but fungal hyphae: filaments far thinner than a hair, threading pores in soil that roots cannot enter. The tree's boundary becomes a negotiated interface, a living membrane where carbon is paid and nitrogen and phosphorus are received. The organism that harvests nutrients is partly fungal. The organism that supplies energy is partly plant. If that fungus also feeds on multiple plants, the interface is not merely between one tree and one fungus. It is between many trees and many fungal partners, each with its own incentives.

Incentives matter because mycorrhizal symbioses are not utopias.

E. Toby Kiers and colleagues demonstrated in 2011 that plants and fungi can discriminate among partners and adjust trade: rewarding those that provide more nutrients and penalizing those that do not. Their experiments, conducted with controlled systems, reframed the symbiosis as a market with bargaining, sanctions, and preferential exchange. This kind of mechanism makes a network plausible without requiring benevolence. A fungus that connects multiple plants may route nutrients toward the plant that pays best. A plant may allocate more carbon to the fungal strain that delivers more phosphorus. The resulting pattern can look, from a distance, like cooperation. Up close, it may be conditional trade.

This is where the solitary-tree story becomes most misleading. Competition certainly exists. Trees compete for light in ways that can be brutal. But belowground, the interactions are not limited to theft and deprivation. They include transactions. They include indirect effects: a tree feeding fungi that change soil chemistry, which in turn changes which seedlings can establish. They include feedback loops: a stand dominated by one species can cultivate a fungal community that favours that species, increasing its dominance. The forest is not a set of isolated bank accounts; it is a tangle of debts, investments, and infrastructure shared with strangers.

One way to see the shift is to revisit the classical forestry idea of *site*. Foresters have long classified land by productivity: the same species grows faster on some sites than on others. Traditionally, this was explained by climate, soil texture, drainage, and nutrient content. All true. But once fungi are recognized as central to nutrient uptake, the site becomes partly biological. Two soils with similar chemistry can behave differently if one has an established mycorrhizal community and the other has been disturbed, sterilized, compacted, or eroded. After a clearcut, after fire, after heavy machinery, the forest is not merely missing trees. It may be missing the underground partnerships that allow seedlings to feed efficiently. A seedling planted into such soil is not a lone pioneer; it is an organism trying to assemble its microbial entourage under hostile conditions.

The historical dogma also shaped how foresters interpreted mixtures of species. In a strict competition framework, mixed stands are puzzles: why would two species coexist if each is trying to maximize its own growth? Ecologists have answers rooted in niche differences and disturbance regimes. Mycorrhizal networks add a subtler possibility: that coexistence can be stabilized or destabilized by shared fungal partners. In some contexts, a network could facilitate establishment by moving limiting nutrients or by buffering seedlings against drought via shared access to soil microsites. In other contexts, networks could *increase* competition by allowing established trees to monopolize fungal pathways and outbid seedlings for nutrients, a kind of underground enclosure. Either way, the solitary-tree myth fails because it assumes that all interactions are direct and local: your shade on me, my root on yours. The fungal web creates interactions that are indirect and extended: your carbon shaping my fungal access, my fungal partners shaping your nutrient supply.

A forest, then, is not merely a density of stems. It is a density of relationships.

There is an emotional temptation here: to swap one moral narrative for another. The old narrative is harsh, almost Victorian: nature as struggle, winners and losers, the invisible hand of competition. The new narrative risks becoming pastoral: nature as mutual aid, elders caring for youth. Both are too tidy. The more interesting reality is that the same mechanism—a shared mycorrhizal network—can produce outcomes that look cooperative in one moment and competitive in another, depending on who is carbon-rich, who is nutrient-limited, who is shaded, who is stressed, and which fungal species are present.

Consider what it means, biologically, for a mature tree to stand in full light while a seedling lives in shade beneath it. The adult produces a surplus of sugar. The seedling is carbon-starved. The adult may be nutrient-limited because its demand is enormous; the seedling may be nutrient-limited because its roots are small. A fungal partner sits between them, hungry for carbon and capable of scavenging nutrients from the soil matrix. If the fungus connects them, carbon can move into the fungal network from the adult. Nutrients can move from the fungal network into both plants. Whether any net carbon ends up in the seedling depends on gradients and controls: source—sink dynamics, fungal storage, and the rules of trade. In Simard's field work, shading shifted net transfer, consistent with these dynamics. That does not prove that forests are guided by generosity. It shows that the solitary-tree model cannot capture the physics of the system, because the plumbing does not respect individual boundaries.

One might object that even if carbon moves, the quantities are trivial—a trace that matters only to scientists with sensitive instruments. This is a fair question, and it has been asked repeatedly, including in a 1999 paper that emphasized how difficult it is to quantify transfer and to demonstrate that it is ecologically meaningful. A molecule can cross a boundary without changing anyone's fate. The burden of proof is not merely movement but consequence: growth, survival, reproduction, resilience. Evidence for large, general effects across forests is mixed and context-dependent. Some experiments suggest that access to established fungal networks can influence seedling performance, especially under stress, while others find limited or inconsistent benefits once alternative pathways are accounted for. The honest posture is neither dismissal nor evangelism. It is curiosity sharpened by skepticism.

Yet even without a universal claim about benefits, the discovery of underground sharing is still intellectually disruptive. The solitary-tree dogma assumes that individuals can be understood by measuring only their own traits and their direct competitors. The fungal perspective insists that a tree is partly defined by its partners and by the network architecture in the soil. A forester can measure diameter and height, crown class and leaf area, and still miss a large part of what determines a tree's health: whether its fine roots are well colonized, which fungal species dominate, whether hyphal cords extend into nutrient-rich microsites, whether the local network is continuous or fragmented.

The myth persists partly because the evidence is literally hard to see. Aboveground, a forest is an architecture of trunks and branches, easy to map, easy to photograph, easy to quantify. Belowground, the relevant structures are delicate and ephemeral. Digging destroys them. Washing roots removes context. Even with microscopy, DNA markers, and isotopes, the work is slow and uncertain. That practical difficulty has shaped the history of ideas. Forestry grew up in the realm of what could be tallied: board-feet, basal area, canopy closure. The fungal web grew in the margins, where the most important structures are measured in micrometres and the most consequential exchanges happen in darkness.

A final image helps replace the solitary tree with something more realistic without making it mystical. Picture a city seen from above at night. Streets glow. Traffic flows. Now picture the same city, but imagine that most of its vital movement happens in tunnels: water mains, sewage lines, power cables, fiber optics. From the surface, you can still model the city as a set of buildings competing for space. You can measure the shadows they cast. But you will misunderstand why some neighborhoods thrive and others decay if you ignore the buried infrastructure that brings in resources and carries away waste. In a forest, the fungal network is not merely a curiosity; it is part of the infrastructure of life, old enough to have shaped the evolution of plants themselves.

Once that is granted, the solitary tree becomes a convenient fiction—useful for some calculations, but too blunt as a worldview. A tree is an individual, yes, but also a node: a photosynthetic engine embedded in a communal medium, connected by fungal threads that can transmit matter and influence in ways that are sometimes subtle and sometimes decisive. The forest is not only competition among trunks in the light. It is also negotiation among roots in the dark, where the boundary of the self is thin, porous, and constantly renegotiated.

1.2 The Crowded Dark

A forest floor looks calm from above: needles and leaves, a soft mat of moss, the occasional fallen branch going grey with age. Step off the trail and the calm becomes a kind of illusion. The ground gives slightly, as if you were walking on a thin skin stretched over something busy. That busyness is not poetic license. Soil is less like a pantry of minerals and more like a city under permanent construction, full of tradespeople, thieves, recyclers, predators, and opportunists. The fungi that link trees live inside this city. They do not run it alone, but they are among its most influential engineers.

The first difficulty in imagining soil life is scale. A tree is large enough to be counted; a fungus is often large enough only to be inferred; the organisms that feed on both can be smaller than a punctuation mark. Yet the numbers are what change the meaning. A handful of forest soil contains thousands of species and an astronomical number of individuals: bacteria, archaea, fungal hyphae, protozoa, nematodes (roundworms), mites, springtails, and the larvae of insects. Some of these organisms can survive long droughts and wake again with the next rain. Some form spores that can wait out winter and wildfire. Many eat each other. Almost all of them, directly or indirectly, depend on carbon that originated in photosynthesis aboveground and sank into the soil as litter, roots, and exudates.

The temptation is to treat this complexity as background texture, the way one might treat the hum of a crowd at a stadium. But the crowd determines the game. In soil, who eats whom and who cooperates with whom decides how quickly dead leaves become plant-available nutrients, how carbon is stored or released as CO_2, how diseases spread, how roots grow, and whether seedlings establish. The mycorrhizal network is not a single telephone line running between trees. It is a set of cables laid into a turbulent marketplace where cables are nibbled, rerouted, and constantly renegotiated.

A simple change of perspective can help. Aboveground, the major food web of a forest is easy to picture: leaves feed caterpillars; caterpillars feed birds; birds feed hawks. Belowground, the dominant flow is less glamorous but more consequential. Dead plant matter and root exudates feed microbes; microbes feed grazers; grazers feed predators; predators excrete nutrients that plants can use. This is sometimes called the *brown food web*, not because it is dull, but because it is built on decay. In many forests, more energy enters the ecosystem through dead organic matter than through herbivory. The forest is fueled not only by what is eaten green but by what is decomposed brown.

Fungi occupy a peculiar position in this economy. Many are decomposers, breaking down tough plant tissues and the complex polymers in wood and leaves. Others form mycorrhizal partnerships with roots, taking sugar from living plants in exchange for nutrients gathered from the soil. In practice these roles can overlap and blur. Some mycorrhizal fungi have enzymes that can liberate nutrients from organic matter; some decomposers can opportunistically associate with roots. What unifies fungi, ecologically, is their form: hyphae, long branching filaments that explore soil like a fine mesh. This form makes fungi both predators and prey. Hyphae can invade and digest microscopic organisms; hyphae can also be grazed by springtails, mites, and nematodes, clipped and consumed like living spaghetti.

Anyone who has tried to keep a compost heap knows how lively decomposition can be. Heat rises from the pile as microbes metabolize. Odors shift as communities change. In a forest, the composting is more diffuse and more disciplined. Leaves fall in autumn and accumulate in layers. Needles and twigs add their own chemistry. Roots continually die and regrow, creating channels and releasing carbon. Fungi and bacteria attack the litter, and in doing so they compete for the same resources. Bacteria generally excel at quick use of simple compounds and can reproduce rapidly; fungi excel at accessing complex substrates and moving resources along hyphae. Neither group is inherently superior; each thrives under particular conditions of moisture, temperature, pH, and litter quality. Their balance matters, because it determines the speed and pathway of nutrient release.

The older story of soil fertility—the one often implied by textbooks and gardening advice—treats nutrients as if they were coins scattered through the soil, waiting to be picked up by roots. In reality, many of the most valuable nutrients are locked in organic forms: nitrogen in proteins, phosphorus in complex molecules, and essential elements bound into dead tissues. Unlocking them requires enzymes and labor, and in soil that labor is performed largely by microbes. Nutrients become available not because they are simply present, but because organisms with particular abilities are present.

This is why mycorrhizal fungi matter even before any talk of tree-to-tree transfer begins. A fine root on its own is a blunt instrument. It can absorb soluble nutrients in the immediate zone around it, but it struggles to scour the soil for scarce phosphorus or nitrogen in forms that are not readily dissolved. Attach that root to a mycorrhizal partner and the effective foraging surface explodes. Hyphae can enter tiny pores and follow microscopic gradients of nutrient concentration. They can proliferate rapidly into patches of rich organic matter. They can also transport nutrients back to the root, turning spatial patchiness into a kind of distributed supply chain.

The crowdedness of the soil also means that mycorrhizal fungi are rarely dealing with a single plant and a single fungus. Roots are colonized by communities of fungi. Fungal hyphae encounter other fungi and bacteria and negotiate boundaries with them, sometimes through direct chemical warfare. A hypha does not extend into empty space; it threads between mineral grains already coated with biofilms. It passes nematodes that probe and graze, protozoa that hunt bacteria, mites that shred litter. In such a world, a fungal network that connects trees is not merely a conduit. It is a living organism trying to survive in a place full of mouths.

A concrete example makes this vivid. Consider springtails, tiny soil-dwelling arthropods that can be so abundant in forest litter that they seem, under a hand lens, like animated dust. Many springtails feed on fungi. They do not politely harvest; they graze, clip, and wound hyphae. The fungus responds by regrowing, thickening, or producing chemical defenses. That grazing, in turn, changes fungal community composition: some fungi are palatable, others are toxic; some grow fast and tolerate damage, others invest in protective structures. The result is that the shape and continuity of mycelial networks are influenced not only by trees and nutrients but by the appetites of small animals you could fit on a fingernail.

The predators of these grazers add another layer. Nematodes, the abundant roundworms of soil, include fungal-feeders, bacterial-feeders, and predators that hunt other nematodes. Protozoa engulf bacteria. Mites prey on nematodes and springtail eggs. Each feeding interaction releases nutrients: grazers excrete nitrogen and phosphorus in forms that plants and microbes can use. In this sense, predation is fertilization. The forest's nutrient economy depends on a cascade of tiny violences and tiny excretions, repeated billions of times.

This matters for the story of fungal networks because it changes the question from *Can trees share?* to *What kind of world would sharing have to occur in?* A hyphal connection between roots is, at minimum, a potential route for movement. But whether resources actually move, how far, and in what direction depends on gradients and constraints. In a crowded soil, gradients are steep and localized. A decaying leaf can create a nutrient hotspot a few millimeters wide. A drying spell can create a mosaic of wet and dry pores. A root can leak sugar and create a plume that attracts bacteria, which attract grazers, which attract predators. The fungal network is embedded in this patchwork, responding to it in real time.

There is also a deeper complication: soil is not just crowded; it is chemically conversational. Roots release compounds that attract mutualists and repel pathogens. Microbes release compounds that stimulate or inhibit root growth. Fungi release compounds that suppress competitors. Some of these compounds are signals—information exchanged between organisms. Others are by-products that nonetheless have effects. Either way, the idea that trees might communicate through fungal networks cannot be evaluated without knowing that the soil is already full of chemical chatter. A molecule passing from one root to another might be a targeted message, a metabolite hitchhiking in water, or a fungal exudate that both roots respond to independently. The soil makes it hard to tell stories with a single protagonist.

A historical anecdote shows how difficult it was to appreciate this. When early plant physiologists first traced carbon and nutrients through plants, they often did so in sterilized or simplified conditions: hydroponic setups, sand cultures, pots of pasteurized soil. The goal was control, and control required removing the crowd. Those experiments produced foundational knowledge about plant nutrition, but they also encouraged a picture of roots as autonomous absorbers. Only later, as ecological methods matured and molecular tools arrived, did it become feasible to study roots in the full riot of living soil. The discoveries about mycorrhizal networks emerged, in part, because scientists began to put the crowd back in the picture.

A striking piece of evidence that fungus-mediated carbon transfer is not merely an experimental artifact comes from plants that have taken the logic of mycorrhizal trade to an extreme. These are the mycoheterotrophs: plants that obtain some or all of their carbon from fungi rather than from photosynthesis. Some are ghostly and pale, lacking chlorophyll, emerging briefly from leaf litter like biological punctuation. They are not cheating photosynthesizers so much as botanical pickpockets, tapping into fungal networks that are themselves connected to green plants. Their existence is a natural proof-of-concept: carbon can move from one photosynthesizing plant into a fungus and from that fungus into another plant in a way that supports the recipient's life. It is difficult to dismiss this as trivial or accidental. A mycoheterotroph is a living receipt.

Of course, a forest is not composed of mycoheterotrophs. Their existence does not prove that most trees routinely donate meaningful carbon to one another. But it does undermine the stubbornest version of the solitary-tree myth: the idea that carbon is strictly private property once fixed by a leaf. In the crowded dark, property lines are porous.

The social life of carbon is only one part of the soil story. Nitrogen and phosphorus, the classic limiting nutrients in many forests, have their own dramas. Nitrogen can be abundant in the atmosphere yet scarce in forms plants can use. Its availability depends on microbial processes: nitrogen fixation, nitrification, denitrification, and mineralization. Phosphorus is often bound tightly to minerals or locked in organic molecules and can be particularly limiting in older, weathered soils. Mycorrhizal fungi have strategies for both: they can forage extensively for scarce phosphorus, and they can intercept nitrogen released by decomposition before it leaches away. When a fungal network moves nutrients, it may be doing so to satisfy its own growth needs as much as the plant's. The fungus is not a courier employed by the tree; it is a partner with its own body to build.

This is where the earlier carbon-transfer experiments acquire their full ecological meaning. If two trees share a fungal partner, that fungus has access to two carbon sources. If one tree is shaded and pays less, the fungus may allocate less nutrient to it. If one tree is stressed and leaks different compounds, the fungal community around it may shift. Source-sink dynamics are not only about carbon; they apply to nutrients and water as well. A shaded tree becomes a sink for carbon; a fast-growing tree becomes a sink for nitrogen and phosphorus; a drought-stressed tree becomes a sink for water. In a connected system, sinks pull.

Yet the crowd insists on humility. A molecule that moves through a fungal hypha is not necessarily moving *for* the benefit of a tree. It may be moving because the fungus is reallocating its own internal stores to where growth is most profitable. It may be moving because grazers have damaged hyphae in one patch, forcing regrowth elsewhere. It may be moving because a bacterial biofilm has altered local chemistry. Even the direction of movement can flip quickly as light conditions change, as observed in the field by Simard and colleagues in the late 1990s. The forest floor is responsive, and responsiveness is not the same thing as intention.

The crowded dark also contains an important check on romantic interpretations: cheating is possible. In any market, some participants try to take more than they give. Plants can allocate carbon to mycorrhizal fungi without receiving proportional nutrients. Fungi can connect to roots and extract sugar while providing little in return. Other microbes can parasitize the interaction, siphoning carbon from exudates and leaving the plant with less bargaining power. The existence of such cheating pressures is one reason the symbiosis has evolved mechanisms of partner choice and sanctions. But the point is broader: the underground network is not a benevolent commons. It is an arena of conflict as well as exchange, and the conflict can be microscopic and relentless.

A walk in an old forest can make this feel less abstract. Turn over a patch of litter and look closely at the underside. You may see white threads, fans, and cords: fungal structures stitching leaves together. You may see tiny pellets of feces from soil fauna, evidence of constant grazing. You may see roots so fine they resemble hairs, coated in a fuzz that is not root hair but fungal mantle. You may smell a sharp, mushroomy odor, geosmin and other compounds released by microbes. That smell is not just pleasant; it is a chemical announcement of metabolism. It is the scent of a crowded place.

To speak of fungal networks coordinating forests, then, is to speak of coordination in a riotous context. The network does not operate in an empty laboratory dish. It operates in soil where every resource is contested, where every filament is at risk of being eaten, where every patch of nutrient is temporary, where chemistry and biology are inseparable. The remarkable thing is not that the underground world is complicated. The remarkable thing is that trees manage to grow tall at all, year after year, supported by partnerships that must be constantly maintained in the face of predation, competition, and decay.

A solitary tree would have to do everything itself: explore soil, defend against pathogens, acquire nutrients from chemically reluctant minerals, cope with drought, and rebuild after damage. In the crowded dark, no organism gets to be solitary for long. The soil does not permit isolation. It presses living things into contact, forces bargains, and punishes those that cannot negotiate. A tree rises from the forest floor like an independent pillar, but its independence is partly an optical illusion produced by distance. Up close, the pillar is threaded into the ground by an intricate mesh of other lives, and those lives do not merely surround it; they help decide what it can become.

1.3 Architecture of the Invisible Highway

Stand in a forest long enough and the aboveground order begins to feel straightforward. Trees sort themselves by height and shade tolerance. Crowns interlock. Branches reach toward gaps. Even the chaos of a windthrow eventually resolves into patterns: a patchwork of light and regeneration. The underground, by contrast, resists the mind's usual tools. It is hidden, soft, and easily destroyed by the very act of inspection. For a long time, the notion that trees might be physically connected belowground was treated as a metaphor or a moral fable. What changed was the realization that the connection is not an idea. It is a structure.

The term *Common Mycorrhizal Network* (CMN) sounds as if it were invented to lend gravitas to a popular phrase. It is, in fact, an attempt to be precise. *Mycorrhiza* means "fungus-root": an intimate partnership in which fungal tissue associates with a plant's fine roots. *Common* signals that the same fungal individual, or at least the same continuous mycelial system, is shared among multiple plants. A CMN, then, is not simply "many mycorrhizas in one place." It is a physical continuity of fungal hyphae that ties multiple root systems into a single fungal body.

That definition immediately raises two questions that are more anatomical than philosophical. What does the network look like in the soil? And how can anyone tell that it is truly *one* network rather than many separate fungal fragments that merely happen to occupy the same patch of earth?

Begin at the root tip. In ectomycorrhizal partnerships, common in many temperate and boreal forests, the fungus forms a sheath (the *mantle*) around the finest roots. Under a microscope the root tip can look as though it has been dipped in felt. From this mantle, the fungus sends hyphae outward into the soil. Some hyphae run as individual threads; others gather into bundles called *rhizomorphs* or *mycelial cords*. These bundles are not mere tangles. In many taxa they are differentiated structures: multiple hyphae aligned, sometimes with specialized channels, creating something more like a cable than a loose net. Their functional promise is obvious. A thin hypha can explore; a thicker cord can transport.

In arbuscular mycorrhizal partnerships—more common in grasslands and in many tropical forests—the physical form differs, with fungal structures penetrating root cells and forming branched arbuscules for exchange. Yet the same basic logic holds: hyphae extend beyond the root's immediate boundary, enlarging the plant's reach into soil. CMNs are discussed most often in the context of ectomycorrhizal systems because the cords and mantles can be conspicuous and because some ectomycorrhizal fungi form extensive, persistent genets that are amenable to mapping.

The word *network* invites us to picture a neat set of junctions and edges, like a subway map. Real soil is messier. Hyphae grow, branch, fuse, and die. They are grazed and regrow. They are interrupted by stones, by dry pores, by hostile bacterial films. Yet the topology—the pattern of connection—is not random. It is shaped by the fungus's evolutionary strategy, by the distribution of resources, and by the presence of hosts.

Reinhard Agerer proposed in 2001 that ectomycorrhizal fungi could be classified by *exploration types*: morphological patterns of extramatrical mycelium (the hyphae extending outward from the root) that reflect how the fungus searches for nutrients. Some fungi are *contact* types, keeping most of their activity close to the root surface. Some are *short-distance* or *medium-distance* explorers, sending hyphae further into the soil but not forming thick cords. Others are *long-distance* explorers, building prominent rhizomorphs that can bridge nutrient-poor zones to reach richer patches. The point of such a classification is not to make soil life sound like military reconnaissance, but to name a real architectural diversity. Two trees in the same stand can be colonized by fungi with radically different construction habits, and thus be plugged into different kinds of underground infrastructure.

This is where the solitary-tree picture collapses at a more technical level. "Roots absorb nutrients" is not wrong, but it is incomplete in the way "cities obtain food" is incomplete. *How* they obtain it matters. A forest can be dominated by fungi that behave like local foragers, working close to home, or by fungi that behave like builders of long cables, moving resources across decimeters and meters. A seedling's prospects may depend not only on how much light reaches its needles but on which exploration types happen to dominate the soil it is planted into.

The physical reality of these structures has always been hard to document, and that difficulty has shaped the history of the idea. Digging up a root system is like trying to map a spiderweb with a shovel. When you excavate soil you break hyphae; when you wash roots you remove the context that tells you where those hyphae led. Even when rhizomorphs are visible, following them through soil is a painstaking exercise in patience and luck. Reviews of ectomycorrhizal ecology have repeatedly emphasized a sobering point: any map we draw of fungal extent is probably an underestimate. The network is almost certainly larger and more continuous than what we can preserve long enough to measure.

So how do you demonstrate a CMN without having to keep every hyphal thread intact?

One answer arrived with molecular genetics. A fungal individual—a genet—has a genetic identity. If you can sample fungal tissue from multiple root tips and show that it shares the same genotype across different trees, you have strong evidence that a single fungal individual colonizes multiple hosts. The network, in this sense, becomes measurable even if you cannot follow every strand physically. You infer the bridge because you can identify the builder.

A striking example comes from work by Kevin Beiler, Daniel Durall, and colleagues (2009), who studied *Rhizopogon* fungi associated with Douglas-fir. Using multilocus microsatellite markers, they matched fungal genotypes across mycorrhizal root tips in a forest plot and inferred a network of shared fungal genets connecting multiple trees, including different age cohorts. Some genets colonized many trees. The resulting structure was not egalitarian: larger, older trees tended to be more connected, functioning as hubs in the sense that they shared fungal genets with many neighbors. "Hub" is a tempting word, because it can smuggle in social meaning, but here it is simply a description of topology: some nodes have more links.

This kind of work did something that earlier tracer experiments could not do on their own. Tracers hinted at resource movement. Genetic mapping showed physical continuity and scale. Together, they made it harder to dismiss underground connectivity as rare or accidental, at least in some systems. The mycorrhizal network was not just plausible plumbing; it was a plausible *architecture*.

Yet even this proof comes with caveats that are worth keeping in view. A shared fungal genotype on two root tips strongly suggests a shared fungal individual, but it does not automatically tell you how continuous the hyphal connections are at every moment. Hyphae can be severed and regrow. A genet can persist in fragments. Soil is dynamic. The network is a living thing, not a static cable. And, crucially, physical connection does not guarantee substantial exchange between plants. A road does not prove trade; it makes trade possible.

The human mind likes to leap from structure to story: if a network exists, it must be used for communication, for help, for coordination. Soil biology is less obliging. A CMN is better understood first as infrastructure for the fungus itself. The fungus benefits from being able to draw carbon from multiple plants, buffer itself against the failure of any single host, and allocate its growth toward the most profitable patches of soil. In this view, the fungus is not the telephone line between trees; it is the company that laid the cable for its own business, and trees are among its clients and suppliers.

Even the most "highway-like" structures, the rhizomorphs, make sense in fungal terms. Building a thick cord is costly, but it can pay off by allowing the fungus to move water and nutrients efficiently, to colonize distant roots, and to persist through dry periods by maintaining continuity between moist microsites. The cord, in effect, is a way of taming the patchiness of soil. It is also more resistant to grazing and physical disturbance than individual hyphae. The highway is not a gift to trees; it is an adaptation to a hostile environment.

A historical anecdote from the laboratory captures how easy it is to underestimate fungal engineering. In controlled experiments, a fungus grown on a petri dish can look delicate, a white fuzz spreading across agar. Transfer it into soil and it becomes harder to see and harder to credit. But in certain conditions fungi reveal their construction in dramatic ways: thick cords running across leaf litter, mycelial fans under bark, dense mats that bind soil. Early forest ecologists who peeled back bark on dead wood and found white sheets of hyphae were often studying decomposers, not mycorrhizal fungi. The lesson still applies. Fungal architecture is often most apparent when it is doing heavy work: invading wood, bridging dry gaps, staking territory. The same kind of work, more subtly, occurs around living roots.

It helps to locate the CMN in a very particular place: the *rhizosphere*. The rhizosphere is the thin zone of soil influenced by roots: enriched by exudates, populated by microbes attracted to those exudates, and chemically distinct from bulk soil. If the soil is a crowded city, the rhizosphere is a busy waterfront where goods arrive, fees are paid, and border controls are enforced. Roots leak sugars and amino acids; bacteria bloom; protozoa graze; nutrients are mineralized and immobilized. Mycorrhizal fungi operate at this interface, negotiating access to carbon on one side and to mineral and organic nutrients on the other.

The physical topology of a CMN can be imagined as a set of root tips, each wrapped in fungal tissue, connected outward through a mesh of hyphae that sometimes bundles into cords. Where the hyphae from one colonized root meet those from another, they may fuse, forming a continuous network. In some fungi, hyphal fusion (anastomosis) allows resources to move across the mycelium as within a single body. In others, the network may be more modular, with less fusion. The details vary by species, and that variation matters. A CMN is not one thing; it is a family resemblance.

The question that often follows is the one that made the "wood-wide web" story famous: can resources move through this network from one tree to another in a way that matters? The cautious answer remains the most useful: movement can occur, and in some field contexts it has been detected with isotopic tracers; the magnitude and ecological consequence vary, and proving the exact pathway is technically difficult. But the architectural point stands regardless of the outcome of any single transfer experiment. Once a fungus links multiple plants, the possibility space of forest interactions changes. A seedling can be connected to the soil not only by its own roots but by the pre-existing fungal infrastructure of its neighbors. A mature tree can influence soil communities not only through its own exudates but through the carbon it supplies to fungi that in turn affect microbial and faunal communities. Disturbance can fragment networks, altering nutrient dynamics in ways that persist even after the aboveground canopy looks restored.

This architectural view also disciplines the imagination. When people speak of "trees talking," it is easy to picture a message intentionally sent from trunk to trunk. A network makes it more interesting and more ambiguous. A fungal cord can transmit not only sugars but also water, ions, and small organic molecules. Some compounds are signals; others are simply cargo. Some effects may be direct transfer; others may be indirect changes in the fungal community that alter how each plant behaves. A shared network can synchronize plants not because they are exchanging whispered warnings, but because they are plugged into the same fluctuating system: the same fungus responding to the same nutrient patch, the same drought pulse, the same grazing pressure.

A final practical implication follows from topology. If some trees are more connected than others—as suggested by genet-mapping studies where larger trees share more links—then removing a tree does not only remove wood and leaves. It may remove a major node in a belowground network, altering pathways and connectivity for neighbors. That does not mean that every large tree is a benevolent matriarch. It means that size and age can correlate with network position, and network position can influence how resources and microbes are distributed in soil. The human intuition that "old growth" feels different may be grounded, in part, in this subterranean continuity: not merely more biomass aboveground, but more established infrastructure below.

The CMN, then, is best approached the way an engineer approaches a bridge: by asking what it is made of, how it is anchored, how it is repaired, and what loads it can carry. Only after that comes the more delicate question of what kinds of traffic move across it, and what those movements mean for the lives rooted at either end.

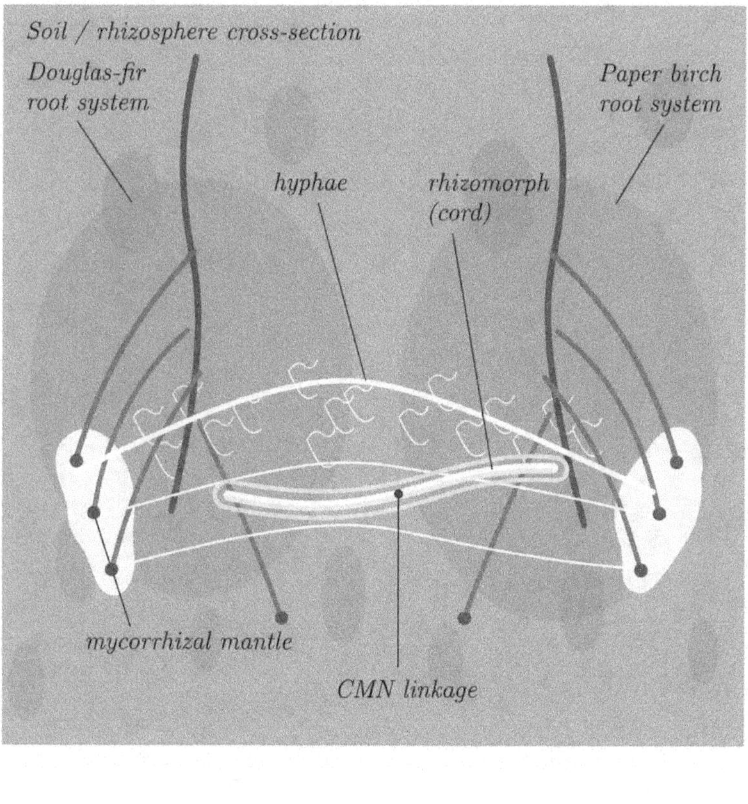

A diagram like this inevitably simplifies. Real roots branch in three dimensions, and fungal networks can connect not two trees but dozens. Still, the simplification is honest in one crucial respect: it shows that the CMN is not an aura or a rumor. It is made of particular tissues in particular places. Root tips are coated. Hyphae extend. Cords form. Connections arise where fungal bodies overlap and fuse. The forest's "internet," where it exists, is not a mist but a material lattice, built filament by filament in the crowded dark, and strong enough to carry the pulse of resources that makes the aboveground forest possible.

Chapter 2

Threads of Life

Beneath the static appearance of a forest lies a dynamic biological manufacturing plant where trees and fungi co-construct a subterranean economy. This chapter dissects the cellular machinery of this partnership, tracing the journey from a microscopic spore to the vast, cable-like infrastructure that binds the woodland together.

2.1 The Subterranean Scaffold

A forest looks like an argument for self-sufficiency. Each trunk seems to rise on its own terms; each crown spreads into its own patch of light. Underground, that impression fails. A tree is not merely *attached* to the soil; it is built into it, and it builds the soil back in return. The roots are the tree's engineering project: a branching framework that must hold the tree upright, harvest water, negotiate toxins, survive droughts and floods, and keep functioning while the world above changes every hour.

Yet even the best-engineered root system has a blind spot. The finest root tips, where most absorption occurs, remain comparatively bulky when set against the clutter of a real soil. Soil is not a smooth sponge. It is a maze of mineral grains, organic crumbs, pores, films of water, air pockets, and bacterial biofilms. Much of the water and many of the nutrients exist not in big, open voids but as thin coatings on particles or within micro-pores that are physically inaccessible to the larger structures of a plant. A root can push between soil aggregates, but it cannot thread itself through every narrow crevice without paying an impossible cost in construction and maintenance. A tree that tried to solve this by making ever finer roots would spend its life converting sugar into tissue, only to have that tissue die back, be grazed, or be cut off by drought.

Root architecture, then, is a scaffold with limits. The tree lays down the beams and pillars, and something else supplies the wiring.

The idea that roots cooperate with fungi is older than most people expect, though it took time to be believed. In the late nineteenth century, when European foresters were struggling to understand why some tree seedlings thrived and others languished in apparently similar soil, a German botanist named Albert Bernhard Frank began to suspect that the pale fungal sheaths he saw on roots were not disease but partnership. In 1885 he coined the term *mycorrhiza*—literally "fungus-root"—to describe a relationship that looked less like infection and more like a negotiated living arrangement. The suggestion was provocative: fungi were widely treated as agents of decay or blight, not as collaborators in building forests. Frank's heresy was that the root was not a closed organ; it was an invitation.

That invitation is written into the way roots are made.

A tree's root system is typically organized as a hierarchy. Thick roots anchor and transport; thinner roots explore; the finest short-lived tips do most of the absorbing. Those tips are not random wisps. They are developmental hotspots, constantly producing new cells and pushing into new patches of soil. This constant renewal is essential because the root tip is where the tree meets the outside world most intimately, and the outside world is rough. Abrasion, herbivory by tiny invertebrates, drying and re-wetting, shifting chemistry: all of it concentrates at the growing edge.

But the very place where a root is most active is also the place where it is most constrained. A root tip needs a protective outer layer, it needs internal plumbing, it needs structural integrity, and it needs to keep its own growth machinery supplied with energy. These requirements set a lower bound on its diameter. Even a very fine root is a complex organ. It cannot become a single-cell thread.

Plants do have an alternative way to increase surface area: root hairs. These are single-cell extensions of the root's epidermis, and they can be extraordinarily numerous. In many plants, root hairs provide a substantial portion of the absorbing surface. They are the plant's attempt to solve the access problem by producing many tiny fingers into the soil's small spaces. But even root hairs have limits. They are short-lived, they require the right moisture conditions to function, and they still cannot match the reach and persistence of a fungal hypha, which is a filament built specifically for penetrating complex substrates.

Here the scaffold metaphor becomes literal. The root system is the durable frame: thicker, longer-lived, more expensive to build. The fungal partner provides a set of finer threads that can be deployed cheaply and dynamically. The tree does not outsource its existence. It outsources a particular kind of exploration.

It is easy to say "fungi are thinner than roots," but scale is where the meaning lives. A hypha can be only a few micrometres across, while a fine root or even a root hair is often an order of magnitude larger. That difference is not merely about size; it is about what spaces are available. Between soil particles there are pore networks that behave like a city's back alleys and service tunnels. A root hair may enter some of them; a hypha can enter many more, and can keep going when a pore narrows, bends, or becomes clogged with organic films.

In drought, the ability to follow thin water films can decide whether a plant stays hydrated or wilts. In nutrient-poor soils, the ability to search widely for rare patches of phosphorus can decide whether a seedling grows or stalls.

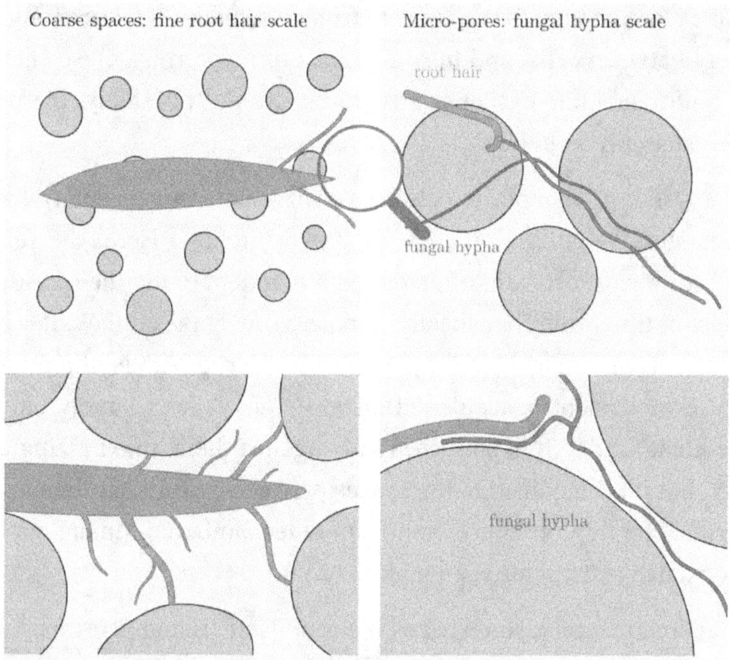

Scale difference: hyphae can penetrate soil micro-pores that root tissues cannot access efficiently.

The scaffold does more than provide a place to attach threads. It sets the terms of trade. Roots are not passive sponges; they are selective, living tissues with immune systems, chemical senses, and strong opinions about who gets close. The surface of a root is a border crossing. When a fungal partner approaches, the root can encourage it with chemical signals, tolerate it, or treat it as an invader. The difference between a pathogen and a partner is not always a matter of species; it can be a matter of context, timing, and negotiation.

Consider the root tip as a zone of risk. If you were a tree, the most dangerous thing you could do would be to open countless microscopic doorways into your tissues without any control. That is essentially what symbiosis requires. The plant must allow a fungus to come close enough to exchange nutrients, sometimes close enough to press between cells, and in some kinds of mycorrhiza close enough to build elaborate exchange structures. To do this safely, the plant uses anatomy as policy.

One striking example is ectomycorrhiza, common in many forest trees such as pines, spruces, oaks, and birches. Here, the fungal partner forms a sheath, often called a mantle, around the outside of the root tip. From that sheath, hyphae grow between the outer root cells to form a latticework known as the Hartig net. Nothing about that arrangement is accidental. The fungus is kept largely outside the plant's cells; it is pressed tightly against them, maximizing contact, but it is contained in the spaces between cells rather than inside them. The tree, in effect, builds an antechamber: intimate enough for exchange, structured enough for control.

That architecture also reshapes the root itself. Ectomycorrhizal root tips often become short and stubby, with fewer or no root hairs, because the fungus has taken over much of the absorptive role. This change can be startling when seen under a hand lens: a normal fine root with hairs looks like a small brush; an ectomycorrhizal tip can look like a tiny club coated in felt. The plant is not being overrun. It is committing to a different design, one that assumes a fungal foraging system will do the exploring in the surrounding soil.

There is an everyday analogy that almost works: it is as if the tree gives up some of its own "fingers" because it has hired a set of far more dexterous gloves. But the analogy fails in one important respect.

Gloves can be removed without consequence. In a mycorrhizal relationship, the root tip is developed in a world where the fungal partner is expected. Remove the fungus, and the root is not simply ungloved; it is under-equipped.

The reason comes back to costs. The currency a tree can pay is carbon, chiefly in the form of sugars made by photosynthesis. Carbon is energy and building material. A tree must allocate it among wood, leaves, defenses, reproduction, storage, and roots. Extending roots through soil is expensive because living tissue must be protected, hydrated, and supplied. Fungal hyphae, though living, can be built and extended with a different cost structure. Their basic unit—a pressurized tube with a wall—is cheap enough to deploy in vast numbers. When a tree trades carbon to a fungus, it is not indulging the fungus; it is purchasing access to a spatial domain the tree cannot economically occupy on its own.

Root architecture makes this trade possible by concentrating exchange at specific points rather than across the entire root system. The finest root tips are the market stalls of this underground economy. They are abundant, distributed through the soil, continually renewed, and close to the resource frontier. Thick roots, by contrast, are highways: durable transport lines that move water and sugars over long distances within the tree. The tree's design separates exploration from distribution, and this separation mirrors the fungus's own division between fine foraging hyphae and more consolidated structures that can move resources efficiently.

Even without naming fungal structures, one can see the logic: a stable scaffold is needed to support a dynamic mesh. A fungal network attached to a plant without a robust transport system would struggle to deliver nutrients to where they are needed. A plant without a fine exploratory partner would struggle to locate diffuse nutrients. The partnership is a coupling of complementary engineering solutions.

The soil environment amplifies this complementarity because nutrients do not behave like water in a pipe. Nitrogen, phosphorus, potassium, and micronutrients are patchy and chemically sticky. Phosphorus, in particular, often binds to mineral surfaces and moves only slowly through soil water. A root that relies on diffusion alone can quickly deplete the immediate zone around it, creating a halo of scarcity. That depleted zone is a common reason plants must keep producing new roots: they are chasing freshness. Fungal hyphae can help by extending beyond the depleted zone, tapping new surfaces, and in many cases releasing compounds that mobilize nutrients from mineral or organic complexes. The details vary among fungi and soils, but the general outcome is that the absorptive surface is not just larger; it is distributed in a way that keeps encountering new micro-sites of opportunity.

This is where the tree's physical presence matters in another way: it stabilizes a long-term relationship in a world of short-term patches. A hypha can race into a nutrient-rich pocket and exploit it, but without a dependable carbon supply it would be forced into a feast-or-famine life. The tree, with its canopy in the light, provides continuity. Season after season, carbon flows down from leaves into roots, and from roots into fungal partners. That steady subsidy makes it worthwhile for fungi to build enduring mycelial systems in soil, and to maintain them even when immediate nutrient returns are low. In a sense, the tree's scaffold turns soil foraging from opportunism into infrastructure.

A seedling makes the stakes visible. A young tree lives close to the edge of starvation. Its first leaves are small; its photosynthetic budget is tight. Its roots are short, and the soil around it may be crowded with competitors. Under these conditions, a successful mycorrhizal partnership can be the difference between establishment and failure. Foresters noticed this long before they knew the mechanisms. When seedlings were transplanted into sterilized or disturbed soils, survival sometimes dropped dramatically. Reintroducing soil from healthy forests could improve establishment, as if the soil carried an invisible ingredient. Often, it did: fungal propagules and the living memory of symbiotic partners.

The root scaffold also mediates conflict. Cooperation in nature is rarely charity; it is managed self-interest. A tree wants nutrients and water at a favorable carbon price. A fungus wants carbon and a stable niche, ideally without being replaced by a cheaper competitor. The root tip, with its regulated entry points and exchange surfaces, is where these preferences collide. Plants can adjust how much carbon they allocate belowground depending on nutrient status and light availability. They can also, to some degree, favor certain fungal partners over others by changing the chemical signals they release or by rewarding more cooperative fungi with more carbon. Fungi, for their part, can channel nutrients preferentially toward roots that pay well. The scaffold is not a neutral platform; it is a bargaining table.

If this sounds too anthropomorphic, it helps to remember that the "bargaining" is accomplished by feedback loops rather than conscious decision. A root cell that receives more phosphorus may change its metabolism and export more sugars; a fungus that receives more sugars may grow more hyphae near that root tip. Over time these local feedbacks can produce patterns that look like strategy. The important point is that architecture enables feedback. Without a defined interface—a place where exchange is concentrated and measured—there would be no reliable way to couple payment to performance.

One can also appreciate the scaffold by imagining failure. When soils are compacted by heavy machinery, pore networks collapse. The large pores that roots use for growth are crushed; oxygen becomes scarce; water movement changes. Roots respond by limiting growth, and fungi respond as well, often with reduced mycorrhizal colonization. In urban trees, where soil compaction and disturbance are common, the loss of a functional scaffold can lead to chronic stress that is visible aboveground as sparse crowns and early leaf drop. The tree may still stand, but the underground collaboration that keeps it supplied is weakened.

Disturbance can also sever the scaffold's finest points. Many mycorrhizal associations concentrate at the most delicate root tips. These tips are constantly being produced and lost. Drought can kill them back; freezing can damage them; browsing organisms can consume them. The tree's investment strategy must therefore anticipate turnover. It builds redundancy not only in branches and leaves but in root tips: many small points of contact rather than a few critical ones. This redundancy is what allows a fungal network to attach at multiple sites, distributing the risk of local damage.

Seen from this angle, "root architecture" is not merely a pattern of branching. It is the physical expression of a tree's expectations about its environment: how stable the soil is, how patchy nutrients are, how intense competition is, how much help is available from fungal partners. Different tree species and different habitats produce different scaffold designs, and those designs filter which fungal partnerships are likely to form. Some roots are built for rapid exploration and quick turnover; others are built for longevity and conservative resource use. The fungal community that assembles around them is not an afterthought; it is constrained and shaped by the scaffold's geometry and chemistry.

There is a quiet intimacy in all this. The part of a tree we rarely see—the pale branching tips, the microscopic hairs, the mucilage that glues soil grains into a sheath, the exudates that feed and recruit microbes—is where the forest's connectivity begins. The trunk may look like the tree's main statement, but the root system is where the tree makes its agreements. It is where it draws a boundary that is also a bridge, sturdy enough to hold fast in storms and subtle enough to admit a partner thinner than any plant tissue could ever afford to be.

2.2 The Infinite Weaver

A root tip is a commitment. It is thick with plumbing, costly to replace, and fussy about where it can go. A fungal hypha is something else entirely: a living filament that can be extended, redirected, abandoned, and multiplied with a speed that makes plant growth look deliberate. If roots are the forest's enduring architecture, fungi are its improvisers—not because they are chaotic, but because their basic unit of construction is so small and so nimble that it can answer opportunities in the soil almost as quickly as those opportunities appear.

The drama begins with a speck.

A spore is a fungal "seed" only in the loosest sense. Seeds are packages of an embryo and lunch. Many fungal spores are more like instructions with a battery: minimal reserves, enough to start, not enough to linger. Spores drift, settle, or hitchhike, and most land in places where nothing happens. The ones that succeed do so because the spore can read its surroundings. Moisture matters first. Without a thin film of water, growth is impossible. Temperature matters. Chemistry matters: a hint of plant root exudates, the scent of decaying wood, the presence of certain bacteria. When the conditions align, the spore opens and produces a single protrusion, a germ tube, like a finger poking out of a glove.

That germ tube becomes a hypha (plural: hyphae), the fundamental building block of a fungus that lives as a network. A hypha is a pressurised cell (or a chain of cells) wrapped in a wall, extending at its tip. Inside, the cytoplasm is crowded with organelles, enzymes, and moving cargo. Outside, the soil is a rough landscape of mineral edges, organic fragments, and microscopic pockets of air and water. The hypha's genius is that it grows at the front rather than swelling everywhere. Most of the action is confined to a tiny region at the apex, while the older parts of the tube remain as a conduit. A fungus can therefore spend its energy on exploration where exploration pays, rather than constantly rebuilding its entire body.

It is worth pausing over the word "body." A familiar animal body is defined by its outline. A fungus refuses that simplicity. The hypha that grows from a spore is not trying to become a compact shape; it is trying to become a set of paths. It extends, branches, and branches again, producing mycelium: the collectively interwoven hyphae that behave as a single organism, able to forage across centimetres, metres, or in some cases much more.

Mycelium looks delicate when it is lifted from soil—a ghostly fuzz, a pale lace—but that impression comes from seeing it out of context. In soil, the mycelium is not hanging in air; it is pressed against particles, threaded through pores, glued into place by secreted compounds, and reinforced where it needs to be. If you have ever pulled apart a decaying log and seen white fans or cords running through it, you have seen mycelium behaving like infrastructure rather than fluff.

How does a hypha extend? The simple answer is that it grows the way a well-managed construction site operates: materials arrive continuously at the expanding edge, while the rest of the structure remains functional. The detailed answer is one of the most beautiful pieces of cellular choreography in biology.

At the tip of many fungal hyphae sits a small, shifting cluster of vesicles and proteins called the *Spitzenkörper*—a German term meaning “apical body.” It is not an organ like a nucleus. It is more like a staging area, a busy roundabout where delivery vehicles merge, turn, and exit. The vesicles carry the building materials of the cell wall and membrane, along with enzymes that remodel those materials as they are laid down. By concentrating this trafficking at the tip, the hypha can extend rapidly and still keep its wall strong enough to resist bursting.

Bursting is not a metaphor. Hyphae are pressurised. They maintain an internal turgor pressure that pushes outward on the wall. Pressure is useful: it provides the force to drive expansion and to nudge between soil particles. Pressure is dangerous: without constant wall maintenance, the cell would rupture. The trick is to keep the wall at the tip temporarily pliable while keeping it sufficiently coherent to hold together, then to stiffen it behind the growing point. The hypha is, in effect, laying down a wall and curing it on the move.

This is why the difference between “growing” and “pushing” matters. A root tip pushes by adding new cells behind a protected cap; it is a wedge. A hypha grows by extending a tube, guided by internal logistics and shaped by the obstacles it meets. When a pore narrows, the hypha can often narrow with it. When a chemical cue appears from one direction, the hypha can bias its growth toward that source. When the path is blocked, the hypha can branch and try alternatives rather than committing its entire structure to a single route. It is a foraging algorithm made of living matter.

Branching is the pivotal act. A single hypha can explore like a probe. A branching mycelium can explore like a search party.

Branches typically form behind the main tip, when a local signal triggers the wall to bud outward and establish a new growing point. Each new tip becomes another site with its own trafficking machinery, its own directional choices, its own capacity to respond to microgradients of nutrients and moisture. A mature mycelium is therefore not one line but a constantly updated map of options: some tips surge forward, others stall, some retreat into dormancy, others reinforce into thicker strands.

The first time one sees a fungus do this under a microscope, it can be disorienting. Early microscopists in the nineteenth century were fascinated by the way fungal filaments seemed to "decide" where to go, even though the language of genes and molecular signalling did not yet exist. Later, as cell biology matured, the Spitzenkörper became one of those discoveries that quietly reorganised an entire field: it offered a physical explanation for how a cell could sustain extreme polarisation—how one tiny region of membrane could behave like the only place that mattered, without the rest of the cell falling apart. That explanation has only deepened with modern work on vesicle trafficking, cytoskeletal tracks, and local recycling at the tip. The hypha is not just an expanding tube; it is a self-organising flow of materials.

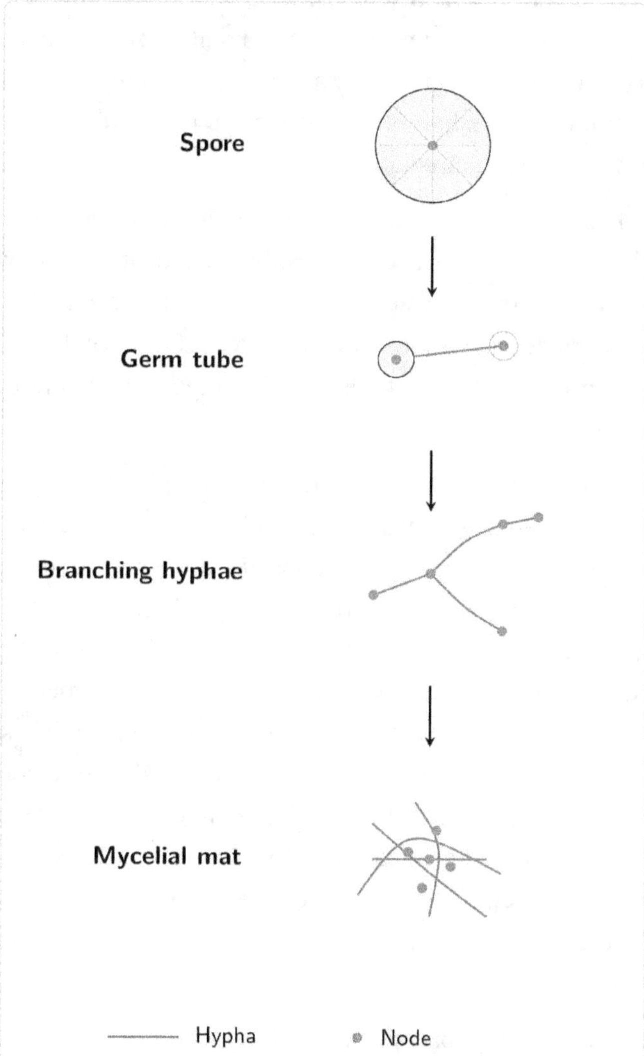

Clean schematic: from a single propagule to a branching mycelium.

The schematic progression from spore to branching mat hides a subtlety that matters in soil: older hyphae are not simply leftover plumbing. They can transport resources. The same tube that once served as a growing edge can later serve as a conduit, moving water and dissolved nutrients from where they are found to where they are needed. A mycelium is not a bag of independent threads; it is a connected system with internal circulation.

The ability to transport changes the meaning of distance. For a root, distance is expensive: to access something ten centimetres away, the plant must build ten centimetres of new tissue and keep it alive. For a fungus, distance can be comparatively cheap if most of the mycelium can remain thin and opportunistic, thickening only where repeated use demands reinforcement.

That reinforcement can be dramatic. Some fungi bundle hyphae into cords or rhizomorphs: thicker, rope-like structures that behave less like a scattering of threads and more like a purpose-built cable. If delicate hyphae are the capillaries of a fungal body, cords are arteries. They are still made of hyphae, but organised: multiple filaments aligned, sometimes with specialised outer layers, sometimes with internal channels that improve long-distance movement. These cords can cross dry gaps, bridge poor substrates, and deliver resources to distant growing fronts. In forests, they help explain an observation that otherwise seems magical: a fungus can exploit a resource patch (a buried log, a cluster of rich litter) and still invest in exploration somewhere else, as if the gains from one place can bankroll the search in another.

A concrete example comes from the fungi that colonise wood. Anyone who has watched a fallen branch turn soft over time has seen fungal power at work, though the mycelium itself is often hidden. Wood is rich in carbon but poor in certain nutrients, especially nitrogen. A fungus feeding on a log is therefore often living in a nutritional imbalance: abundant energy, limited protein-building material. A cord-forming fungus can respond by sending out exploratory hyphae and cords into the surrounding soil to locate nitrogen-rich patches, then bringing that nitrogen back to the wood, where it can be used to build enzymes and more tissue. The log becomes not merely food but a base camp.

The same logic applies when a fungus is partnered with a tree. Carbon may be plentiful near roots. Minerals may be scarce. The fungus can specialise in the search for scarce goods, using its growth form as an adaptive advantage. The mycelium's map of the soil is written not on paper but in the distribution of tips: where they proliferate, where they stall, where they fuse, where they retreat.

The phrase "ability to span vast distances" can be misunderstood. It does not mean a single hypha simply grows straight for metres like a fishing line. In real soil, paths are interrupted. Predators graze. Dry spells create barriers. The mycelium spans distance the way a city spans distance: by building many streets and side streets, accepting local damage, rerouting, and maintaining enough redundancy that the whole does not fail when a part is cut.

One of the best-known demonstrations of fungal scale is also one of the most easily sensationalised: the enormous genetic individuals discovered in certain fungi, where what appears aboveground as scattered mushrooms can be the fruiting bodies of a single organism extending through the soil. The headlines often imply a monstrous blob. The reality is more interesting. The organism is large because its growth mode makes largeness possible without central planning. A hypha does not need to know the size of the whole. It needs only to keep building at the tip, branching when conditions encourage it, and maintaining internal continuity well enough that resources can be shared. Largeness emerges as a by-product of a simple rule: keep exploring, keep connecting, keep feeding the active edges.

This growth mode carries its own kind of vulnerability. Because the hypha extends at a tip, tip damage matters. A soil mite that nips the apex can halt a line of growth. A drying event can kill exposed threads. Yet the mycelium's response to vulnerability is multiplication. Branching is insurance. If one tip dies, another nearby may still be active. If one route dries, another route may still carry water films. The mycelium does not bet everything on a single front; it behaves like a distributed search.

Tip growth also makes fungi exquisitely responsive to fine-scale gradients. A plant root can sense nutrients and steer growth, but its steering is constrained by the physics of multicellular tissue. A hypha's steering is local and continuous. If the concentration of an attractive compound is slightly higher to the left, the delivery of wall-building vesicles can be biased to the left, and the hypha will curve. These small curvatures accumulate into complex trajectories. Over time, the mycelium becomes a physical record of chemical landscapes that no human eye ever sees directly.

There is an emotional temptation to read intention into this. The hypha appears to "seek." It appears to "avoid." It appears to "choose." Better language is available: it *responds*. The response is still astonishing because it is accomplished without nerves or muscles, purely through growth dynamics. The hypha's behaviour is a kind of slow motion, embodied computation: decisions made by distributing building materials in space.

This embodied computation has a cost: the fungus is always building itself into the world. It cannot detach and walk away. When it meets an obstacle, it must either push, go around, or branch and try again. When conditions worsen, it can slow growth or form more resistant structures, but it remains fundamentally a creature of extension. Even retreat is often a kind of reallocation: withdrawing resources from older regions, abandoning unprofitable threads, reinforcing routes that still pay.

The result is a living mesh that can be simultaneously fragile and persistent. Fragile, because any single thread is easy to break. Persistent, because the mesh as a whole can endure breakage. That paradox is the signature of the fungal form.

In a handful of damp soil from a garden, there may be kilometres of hyphae, most of them too thin to see without magnification. That claim sounds like a party trick until one remembers what "kilometres" means at a micrometre scale. Length is cheap when diameter is tiny. And when length is cheap, contact becomes abundant: more contact with mineral surfaces, more contact with organic debris, more opportunities to encounter roots, more chances to form partnerships, more ways to move resources from one micro-site to another.

Fungal mycelium is often described as "threadlike," which is true but incomplete. Thread suggests passivity, something spun and then left. A hypha is an active frontier: it builds itself as it goes, and it changes the soil as it does so. It releases enzymes that break down complex molecules; it exudes compounds that alter pH and mineral availability; it lays down sticky substances that help bind particles into aggregates. The fungus is not merely in the soil. It is one of the forces that makes soil behave like soil rather than like ground rock.

At the tip, all of this power is concentrated into a point so small it would fit comfortably across the width of a human hair many times over. The mycelium's vastness comes from repeating that point, multiplying it, and keeping enough internal coordination that the scattered points add up to something coherent: a living system that can forage, transport, and persist, all while remaining mostly invisible. In a forest, the mushrooms are the punctuation marks. The real sentence is written underground, in hyphae that extend, branch, and braid themselves through the world's narrowest spaces.

2.3 Fusing the Circuit

A hypha can be brilliant at exploration and still be lonely. A single filament, no matter how deftly it curves through pores, is a line without backup. A root tip, no matter how generously it offers sugars, is a single customer at the edge of an enormous market. The forest-wide web that people imagine requires something more specific than "fungus meets tree." It requires contact points that stay in place long enough to matter, and it requires connections between fungal threads that turn many separate explorations into one coordinated body.

The physical formation of that web is less like draping a net over roots and more like electrical work: insulation must be stripped in a controlled way, junctions must be soldered without shorting, and circuit breakers must remain ready to snap shut when the wrong connection is attempted.

The first step is attachment, and attachment is not trivial. The soil surface of a root is not bare; it is coated in mucilage, bathed in a thin chemistry of exudates, and defended by living cells that can recognize trouble. Fungi approach with their own toolkit: enzymes that can soften surfaces, adhesive compounds, and growth patterns that allow them to press close without being crushed. When the partnership forms, the root does not merely tolerate the fungus; it reshapes itself to make a stable interface.

Two broad designs dominate forest stories.

In ectomycorrhiza (common on many temperate and boreal trees), the fungus wraps the root tip in a mantle and grows between outer root cells, forming the Hartig net, a labyrinthine lattice that maximizes contact without entering the plant's cells. In arbuscular mycorrhiza (widespread across land plants, and often considered ancestral), the fungus goes further: it enters root cortical cells and builds delicate, tree-like structures called arbuscules. These are not little fungal trees for decoration; they are exchange organs, highly branched to increase surface area, with the plant building a specialized membrane around them so that the two partners can trade across a controlled boundary rather than mixing cytoplasm.

Both designs solve the same problem: how do you get close enough to exchange nutrients and carbon without dissolving the distinction between "self" and "other"? That distinction matters because fungi are not pets. They are opportunists in the best biological sense: relentlessly responsive to resources. A fungus that could simply consume a root from the inside would do so unless the plant imposed rules. A plant that could take fungal nutrients without paying carbon would do so unless the fungus imposed rules. The interface is the treaty text, written in anatomy.

A treaty, however, is not a network. A network begins when the same fungus connects more than one root, or when a fungal colony grows large enough that resources found in one patch of soil can support growth elsewhere. That is where the second step becomes decisive: hyphae must connect to other hyphae.

This connection is called anastomosis, a term borrowed from anatomy, where it describes the joining of blood vessels. In fungi, anastomosis is the fusion of hyphal threads. Two tips touch, their walls soften at the contact point, membranes merge, and a bridge opens between the cytoplasms. Suddenly, what were two separate lines become one continuous pathway. Water, nutrients, signals, even nuclei in some cases can move across that bridge.

At a human scale, it is hard to see why fusion matters. Aren't there already countless hyphae in a mycelium? But the difference between "many filaments" and "many filaments that can share" is the difference between a crowd and an organization. Without fusion, a fungus is forced to treat each branch as an island. With fusion, it gains redundancy: alternative routes around damage, loops that maintain flow if one segment is severed, and the ability to redistribute resources from older, richer zones to new growing fronts.

Anastomosis also changes the meaning of injury. Soil is violent at microscopic scales. A passing springtail can shear threads; drying can collapse water films; a competing fungus can chemically scorch a boundary. A purely branching system, without fusions, is a tree-shaped graph: break a main branch and everything beyond it is cut off. Add fusions and the graph gains cycles: there are multiple ways to reach the same destination. The mycelium becomes more like a city grid than a set of cul-de-sacs.

Hyphae approach and recognize contact — Anastomosis creates a cytoplasmic bridge

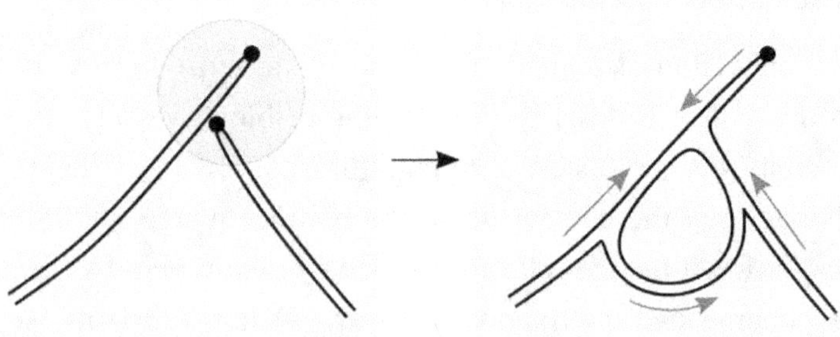

Two tips align; walls locally soften. — Fusion adds redundancy and enables internal transport.

Fusion sounds cooperative, almost tender, until one remembers what it enables. A fused network can share benefits, but it can also share harm. If one region of a mycelium becomes infected by a virus-like element, invaded by a parasitic microbe, or flooded with toxins, connectivity can spread the problem. Many fungi therefore treat fusion as a conditional privilege. They have genetic systems of vegetative incompatibility—often described in terms of *het* (heterokaryon incompatibility) loci—that function like identity checks. When genetically different individuals attempt to fuse, the result may be rapid compartmentalization, plugging of the connection, and even programmed cell death in the fused region. It is a biological circuit breaker: damage is contained by sacrificing a small piece of tissue.

This matters for how "wood-wide webs" should be imagined. Popular accounts sometimes picture a benevolent communal network through which resources flow freely from tree to tree, as if the forest were one family sharing a pantry. Real fungal networks are more guarded. Even within a single fungal species, not every encounter is a handshake. Fusion may occur readily within one mycelium and be blocked between neighbours. The forest underground is full of boundaries.

Those boundaries are not only fungal. Plants add their own. Even when two trees are linked by a "common mycorrhizal network"—a shared fungal mycelium connecting roots of multiple plants—the existence of a physical link does not guarantee an egalitarian flow of resources. Flows depend on gradients and demand. Carbon tends to move from sources (well-lit leaves) to sinks (shaded tissues, growing tips, storage). Nutrients move from patches where they are acquired to places that can use them. A mycelium can be the pathway, but it is not the only pathway, and it is not always the dominant one.

A famous set of experiments helped bring these possibilities into focus. Suzanne Simard and David Perry (1997), working with field systems in North American forests, used isotope labelling to track carbon movement between paper birch (*Betula papyrifera*) and Douglas-fir (*Pseudotsuga menziesii*). Their results suggested that carbon could move in both directions and that net transfer depended on conditions such as shading, which shifts which tree is a stronger carbon source and which is a stronger sink. The work was compelling because it was rooted in real forests rather than pots, and because it treated the underground not as a black box but as a set of measurable pathways.

Over the years, the phrase "common mycorrhizal network" became a banner under which many claims marched: seedling rescue, parental care, forest cooperation, even a kind of vegetal internet. Some of these claims may be true in particular contexts; many are harder to support rigorously. Justine Karst, Melanie Jones and colleagues (2023) argued that parts of the literature on common mycorrhizal networks in forests have suffered from overinterpretation and positive citation bias, and they emphasized how variable field results can be, and how often alternative explanations exist. A molecule found in a neighbour's tissues may have arrived by roots mingling in soil water, by microbes, by mycorrhizal hyphae, or by a mixture of all three. Untangling these routes is not impossible, but it is technically demanding, and forests are noisy systems.

This is not a disappointment; it is a reality check that makes the biology more interesting. A network that is sometimes a conduit and sometimes merely a proximity effect demands better questions. Under what conditions does a fungal pathway dominate? How quickly can material move along hyphae compared with diffusion through soil water? When a seedling benefits near an older tree, is the benefit carbon, nitrogen, water, protection from pathogens, or a change in the local microbial community? The most honest answer, often, is: it depends on the species involved, the season, the soil, the light regime, and the identity of the fungus.

Still, the physical fact remains: fungi create the possibility of connection, and that possibility is built by colonization and fusion.

Colonization itself can be surprisingly intimate and surprisingly constrained. In ectomycorrhiza, the fungus invests in a mantle that can be thick and continuous, a living sleeve. The sleeve protects the delicate root tip, changes its chemical environment, and provides the launching pad for extramatrical hyphae that extend into soil. The Hartig net, threaded between plant cells, is where the closest bargaining happens: sugars moving out of the plant, mineral nutrients moving in. Yet even here, the plant often mounts a kind of measured defensive posture. Work on birch roots by L. Feugey and D.-G. Strullu (1999) described transient defence-like responses during Hartig net formation, consistent with the idea that partnership is negotiated at a guarded border. The tree is not capitulating; it is controlling where the fungus may go.

In arbuscular mycorrhiza, the intimacy goes deeper. The fungus enters cells, but the plant insists on keeping it at arm's length by building a membrane around the arbuscule. The resulting interface is almost baroque in its precision: a fungal structure ramifying inside a plant cell, yet separated by plant-derived membrane and a narrow space where exchange occurs. The fungus gains a private room; the plant keeps the key.

Once multiple roots are colonized, and once the mycelium has enough internal connectivity, the system behaves like a circuit in the ordinary physical sense: there are paths of low resistance along which flux is easier. These paths can change over time. A hyphal bridge that is useful becomes reinforced by repeated traffic; an unprofitable region is abandoned. When damage occurs, detours are explored, and fusion can stitch together new routes. In some fungi that form cords or rhizomorphs, those macroscopic strands can themselves fuse, creating loops that make the transport system more robust. Mapping work on fungi such as *Armillaria* has shown how such networks can develop complex topologies, with some connections acting as critical bridges whose loss fragments the system, and others contributing to resilience by offering multiple alternatives.

The forest floor is full of tiny negotiations of this kind. A fresh root tip appears in spring, exuding sugars and amino acids. Hyphae converge, some welcomed, some resisted. A compatible thread fuses into an existing mycelium, and the new tip becomes another node in the network. Elsewhere, two hyphae meet and fail the genetic handshake, sealing off a risky connection before it can become a highway for trouble. Elsewhere still, a fungus begins life from a cluster of spores, and a special kind of fusion during early colony formation—through conidial anastomosis tubes, in fungi that make them—allows individuals to join briefly before incompatibility systems clamp down later, a short window in which novelty can be assembled from multiple starting points.

It is tempting to treat this as an abstract story about connectivity. It is also a story about very practical outcomes. Whether a seedling survives a dry week, whether a mature tree withstands nutrient stress, whether a patch of forest regenerates after disturbance: these events can turn on microscopic bridges that either exist or do not. The bridges are built from cell walls and membranes, from signalling cascades and recognition proteins, from the gritty mechanics of tip growth and the delicate chemistry of mutual tolerance. A network is not declared into existence. It is fused, junction by junction, in the dark, in spaces too small for roots, by organisms that make connection possible while never forgetting the cost of connecting.

Chapter 3

The Ancient Bazaar

Beneath the forest floor lies a bustling marketplace where carbon is the global currency and phosphorus the precious metal. Here, distinct fungal guilds construct elaborate trading posts within plant roots, negotiating exchange rates that shift ruthlessly between mutual benefit and opportunistic theft depending on the supply of local resources.

3.1 Architectures of Intimacy

Walk into an old-growth forest and the trunks insist on individuality: each tree a separate column of wood, each crown staking out its patch of light. Underground, the story is less solitary. The business end of a plant is not the green canopy but the living skin of its roots, and that skin is almost never alone. Fungi meet roots not as occasional visitors but as long-term partners, settling into a physical arrangement so intimate that it becomes hard to say where plant ends and fungus begins.

Two designs dominate this underground companionship. They are not just two "types of mycorrhiza" (the word for fungus—root partnerships), but two distinct solutions to the same problem: how to exchange goods without being eaten, enslaved, or ignored. Arbuscular mycorrhizae, usually abbreviated AM, are the older arrangement in evolutionary terms and the most widespread across plant lineages. Ectomycorrhizae, EM, are comparatively younger and are concentrated in the great woody families that structure many of the world's forests. Their differences are not superficial. The very architecture of their contact with roots dictates what they can trade, how they bargain, and how widely their influence spreads.

AM fungi behave like meticulous interior designers. They enter the root and build ornate structures *inside* living cells. EM fungi act more like tailors, wrapping the root in a fitted fungal coat and threading their hyphae *between* cells, never pushing through the plant's cell walls. Both strategies create an exchange surface where plant carbon (energy-rich sugars and fats) is swapped for fungal-delivered nutrients and water. Yet the physical route to that surface shapes everything: the speed and control of exchange, the way either partner can regulate the interface, and the kinds of plants that can comfortably host the relationship.

Start with the AM partnership, because it is so common it can feel like background noise. Most herbaceous plants you know—grasses in a lawn, wheat in a field, clover in a roadside verge—make AM. So do many trees, especially in tropical forests. The fungus belongs to a lineage called the Glomeromycota, famous among mycologists for an almost stubborn dependence on plants: these fungi are "obligate biotrophs," meaning they cannot complete their life cycle without living plant hosts providing carbon. That dependence has consequences. If you must live off a partner, you need access to their living tissues, but you must reach them without killing them. AM fungi solve this by creating a controlled, membrane-separated intimacy inside root cortical cells.

The signature structure is the *arbuscule*, named for its resemblance to a tiny tree. A thin fungal hypha pushes into the root, navigates through the outer tissues, and then enters a cortical cell. Crucially, it does not rupture the plant's plasma membrane. Instead, the plant envelops the invading hypha with a new sheath of membrane, continuous with the plant's own membrane but specialized for symbiosis. The fungus branches repeatedly within this compartment, producing the arbuscule's elaborate, coral-like form. The result is a labyrinth of fungal surface pressed close to a plant-derived membrane, like two hands clasped through a thin glove. In biology, surface area is opportunity: every extra fold is another place to move molecules across.

This "gloved handshake" matters because it lets the plant control access. The plant can pack the periarbuscular membrane with transport proteins tuned to the symbiosis. A well-known example comes from *Medicago truncatula*, a small legume that became a model system for studying mycorrhiza. In the early 2000s, Matthew Harrison's group identified a phosphate transporter (often referred to as MtPT4) that is switched on specifically in arbuscule-containing cells, and localizes to that symbiotic membrane. The point is not the gene name, which is parochial, but the principle: the plant builds a dedicated import doorway where the fungus is offering its goods. This is not passive seepage. It is infrastructure.

Ectomycorrhizae take a different route to intimacy, and the difference is visible even to the naked eye if you know what to look for. In a pine forest, pull up a fine root tip and you may see it looks swollen, forked, sometimes even tinted. That altered root tip is often an EM root: the plant's own tissues remodeled under fungal influence. EM fungi come from several fungal groups, most famously the Basidiomycota (the lineage that includes many mushrooms) and also many Ascomycota. Unlike AM fungi, many EM fungi can live in soil without an active host for at least part of their life, and many can deploy powerful enzymes outside the root. In exchange for carbon, they are often formidable scavengers of nitrogen and phosphorus in complex soils.

Their architecture reflects a different kind of negotiation. Instead of entering root cells, the fungus forms a dense outer sheath called the *mantle* around the root tip, sometimes several hyphal layers thick. From this mantle, hyphae penetrate inward but stop at the cell wall. They weave between the root's outer cells, filling the spaces like mortar between stones. This intercellular labyrinth is the *Hartig net*, named after the nineteenth-century botanist Robert Hartig, who described these structures while studying tree diseases and wood decay. The Hartig net is the exchange interface: fungal hyphae on one side, plant cell walls and membranes on the other, with no breach into the living cell interior.

The contrast between AM and EM is sometimes described in a single sentence: AM goes *through* the root's living cell layer (without breaking the plant membrane), while EM goes *around* it. But the consequences deserve a slower look. An arbuscule is a private booth built inside a cell, one fungus at a time, parceled into thousands of tiny trading posts along a root. A Hartig net is more like a corridor network between cells, spreading along the root tip's outer layers and connecting to a mantle that can coordinate activity across the whole tip. If AM suggests countless small deals in many separate rooms, EM suggests a negotiated contract at a shared border crossing.

The nineteenth century provides a useful anecdote because it captures how surprising these arrangements once were. When the German botanist Albert Bernhard Frank introduced the term "mycorrhiza" in 1885, he was trying to solve a practical mystery: why did certain forest trees fail to thrive when transplanted, even when watered and apparently healthy? Frank argued that the fungus on their roots was not a disease but an essential partner. This was a radical claim in an era when fungi were mostly framed as rot, blight, and spoilage. Frank's insistence that trees had *helpers* in the soil was not a sentimental gesture toward harmony; it was an attempt to explain a real-world forestry problem. The architectural features we now teach as textbook distinctions—mantles, Hartig nets, arbuscules—were, at the time, evidence in an argument about whether a fungus could be something other than an enemy.

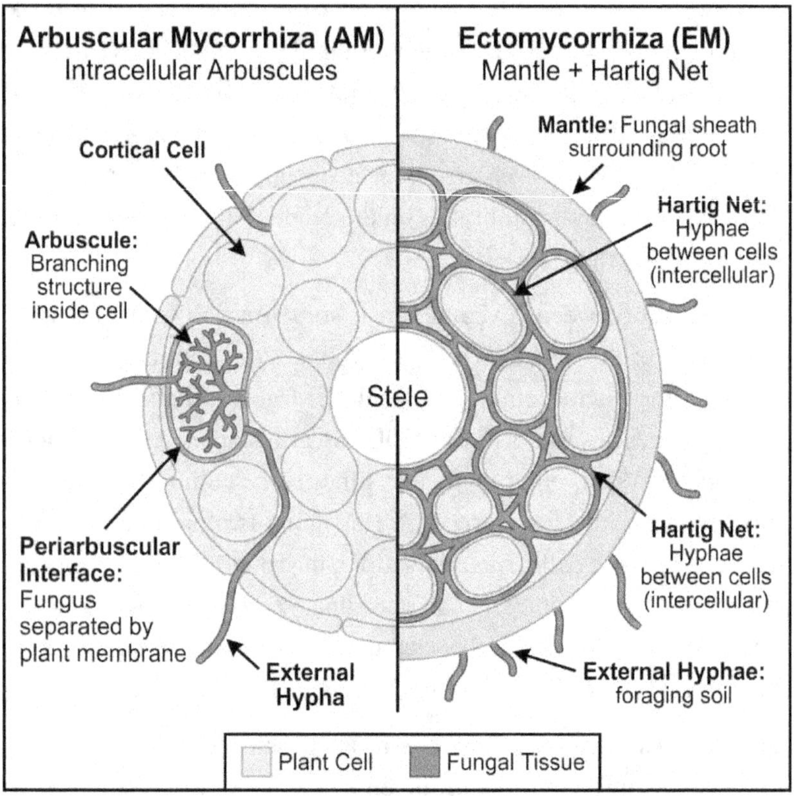

These architectures are not mere anatomical trivia. They are ways of setting the terms of trade. Consider control first. In AM, every arbuscule is temporary. The fungus builds it, the plant tolerates it, and then, after days to weeks, the plant dismantles it and the cell returns to a more ordinary state. The short lifespan is sometimes treated as housekeeping, but it can also be read as governance. By constantly rebuilding the interface, the plant keeps the relationship on a short lease. The fungus, meanwhile, keeps its foothold by colonizing new cells and maintaining an external network in the soil that can quickly supply nutrients when the plant's demand rises.

EM roots, in contrast, are remodeled organs. The mantle can persist as long as the root tip remains active, and the plant often alters its own branching pattern, producing dense clusters of short roots that present many tips for fungal coats. If AM resembles a set of micro-contracts signed and renewed in many individual rooms, EM resembles a recurring agreement attached to particular root tips, each tip becoming a dedicated trading hub.

The architectures also influence *who* tends to partner with whom. AM is nearly universal among herbaceous plants and many tropical trees, which makes sense for a symbiosis that is ancient and closely tied to the basic problem of getting phosphorus out of soil. Phosphorus is often scarce in a form plants can grab quickly, because it binds tightly to minerals. AM fungi are exceptionally good at exploring tiny pores and micro-sites where phosphate is available, then delivering it to roots. Their fine hyphae extend the plant's reach beyond the zone around each root where nutrients have been depleted, but the more distinctive feature is the controlled interface inside the cell, which lets the plant pull phosphate in efficiently when it is offered.

EM partnerships, concentrated in conifers (pines, spruces, firs), many hardwoods (oaks, beeches, birches), and some iconic southern-hemisphere families (such as eucalypts and southern beeches), match a different ecological niche. Forest soils, especially in cool climates, often lock nitrogen inside complex organic matter. Many EM fungi can access that nitrogen by secreting enzymes and acids into the soil, loosening nutrients from stubborn compounds. That kind of chemical work is expensive and risky. It helps to have a root tip that is more like a fortified base—a mantle-wrapped structure that can support and coordinate a larger fungal apparatus outside.

A concrete example makes the contrast less abstract. Take a pine seedling colonized by an EM fungus in the genus *Suillus*, the kind of association that produces the slippery-jack mushrooms familiar to foragers. The pine invests in a cluster of short, mantle-covered tips. From each tip, *Suillus* sends out hyphae that can spread through the litter and the mineral soil, encountering patches of organic nitrogen and pockets of mineral nutrients. The mantle and Hartig net create a wide, shared interface; the fungus can deliver nitrogenous compounds and phosphate, and the tree can deliver carbon in return. The whole arrangement suits the life of a long-lived tree that must manage nutrients over decades.

Now compare that to a clover plant in a meadow making AM with a glomeromycete fungus. Clover roots are fine, numerous, and short-lived. The AM fungus builds arbuscules inside cortical cells as the root grows, each arbuscule a high-surface-area exchange structure tuned for phosphate uptake. The plant's side of the interface is exquisitely molecular: transporters in the periarbuscular membrane, pumps to maintain gradients, enzymes to process incoming nutrients. The fungus's side is architectural: branching hyphae pushed into a tight space that increases contact and speeds exchange. For a plant living fast and responding quickly to weather, grazing, and competition, a symbiosis made of many small, replaceable trading posts is a sensible design.

Neither architecture is "more advanced." Each carries vulnerabilities, and each invites particular kinds of conflict. Intracellular intimacy, as in AM, risks intrusion: if the fungus pushes too hard, it could become a pathogen. The plant's answer is membrane control and timed dismantling. External sheathing, as in EM, risks enclosure: a thick mantle can potentially divert plant carbon into fungal biomass without paying back enough nutrients. The plant's answer is to regulate carbon flow into the tip and to adjust how many root tips it allows to be colonized in a given soil condition. The point is not that plants are wise accountants, but that evolution has built levers into the architecture. When you see the interface, you can often guess where the levers are.

These designs also scale differently. AM fungi form networks in soil that can connect multiple plants, sometimes building what ecologists call common mycorrhizal networks. But because AM exchange happens in discrete arbuscules inside many individual cells, the relationship often feels "cellular" and fine-grained: a mosaic of tiny interfaces that can be turned on or off locally. EM exchange, centered in mantle-covered root tips, can be "organ-level," with a whole tip acting as a unit. That difference may help explain why EM fungi are so often associated with mushrooms that appear above ground: large, organized bodies that mirror, in their own way, the organized hubs they build below.

If the underground bazaar has stalls and gates, then AM and EM are two different ways of deciding where to put the door. AM brings the trading floor indoors, setting up countless small counters inside living rooms, each separated by a membrane the host controls. EM builds a border market at the edge of the root, wrapping the host in fungal cloth and laying out a corridor system between cells where goods can change hands. Both are intimate. Both are risky. Both are, in their own architectural language, contracts written in living tissue and enforced by the geometry of contact.

3.2 Mining the Indigestible

A root is an organ designed for reach, but reach is not the same as access. Even when soil tests declare a field "rich," plants can behave as if they are hungry, not because atoms are absent but because they are locked away in forms roots cannot easily take. Much of plant nutrition is a problem of chemistry and scale: nutrients are either stuck to mineral surfaces, stitched into complex organic molecules, or stranded behind a physical barrier too small for a root hair to enter. Fungal hyphae—the threadlike filaments that make up most of a fungus's body—change the terms of the problem. They are finer than the finest root hairs, able to slip into microscopic pores, and they bring with them a biochemical toolkit that roots largely lack.

The most immediate obstacle is the *depletion zone*, the thin halo of soil around a root where easily available nutrients have already been taken up. Imagine sprinkling sugar on a wet tablecloth. Close to your fingers, the crystals dissolve and vanish quickly; farther away, they linger. Roots create the same pattern with phosphate ions, ammonium, nitrate, potassium, and a host of micronutrients. As the plant absorbs, diffusion tries to refill the area, but diffusion in soil is slow and irregular. A root can sit in a pocket of apparent plenty and still experience local famine.

Hyphae are built for this mismatch between abundance and accessibility. A single filament can extend far beyond the depletion zone, exploring soil volumes a root would need weeks to reach, and because hyphae are narrow, the cost of building that reach (in carbon supplied by the plant, or gathered by the fungus in other contexts) can be relatively small compared to building new root tissue. But reach alone would only bring the fungus to more of the same locked pantry. The deeper advantage is that many mycorrhizal fungi can *alter* the pantry: they can loosen phosphate from mineral surfaces, pry nitrogen from dead organic matter, and trade those liberated nutrients back to plants at the intimate interfaces described earlier.

Phosphorus is the classic case. Plants need it for ATP (the cell's energy currency), for DNA and RNA, for membranes, for almost everything that makes growth possible. Yet phosphate in soil has a frustrating habit: it binds to iron and aluminum oxides in acidic soils and to calcium compounds in alkaline soils. The phosphate may be present in grams per kilogram of soil, but only a tiny fraction is in the soil solution at any moment, and roots can only absorb what is dissolved. This is why phosphorus fertilizer can be both essential and wasteful: much of what is applied becomes fixed again, returning to the locked form.

Fungi attack the problem from several angles. Their hyphae physically extend the absorbing surface area of the plant, but they also secrete organic acids and enzymes, changing pH locally and mobilizing phosphate. Some fungi release phosphatases, enzymes that clip phosphate groups from organic molecules. Others release low-molecular-weight compounds that compete for binding sites on minerals, freeing phosphate into solution. The key is localization. A hypha does not have to change the chemistry of an entire field; it can work a mineral surface micrometers away, where the released phosphate is immediately captured and moved along the fungal network before it is re-adsorbed.

Nitrogen poses a different challenge. Plants can absorb nitrate and ammonium readily, but in many forest soils much of the nitrogen is tied up in complex organic matter—proteins, chitin, and the dark, stubborn mixture often called humus. Unlocking that nitrogen requires extracellular enzymes. Many ectomycorrhizal fungi are particularly well equipped here. Some lineages, such as species in the genus *Cortinarius*, have been linked through field studies and genomic evidence to oxidative enzyme systems that can attack complex organic compounds, including manganese peroxidases associated with the oxidation of humus. The phrase "mining" fits because the process is not gentle scavenging. It is chemical work that alters the soil's most resistant carbon pools, releasing bound nitrogen (and often phosphorus) in the process.

This is not simply the fungus doing the plant a favor. Enzymes are expensive to make, and the products of decomposition are valuable not only to the plant but to the fungus itself. When a mycorrhizal fungus invests in a patch of organic matter, it is prospecting for its own growth and reproduction. The plant's leverage lies in carbon. The fungus's leverage lies in access to nutrients. Each partner specializes in something the other cannot do efficiently. The partnership makes economic sense in the bluntest biological way: it increases fitness under many conditions. But the "market" only exists because the fungus can reach the locked forms and then convert them into transportable currency.

A historical detour sharpens what is distinctive about fungal mining. Early agronomists understood, long before molecular biology, that crop rotations and manures restored fertility, and that legumes improved soils. They had names for "humus" and "mould" and "muck," but the idea that invisible organisms actively regulated nutrient availability was slow to harden into a scientific fact. In the late nineteenth and early twentieth centuries, soil science was transformed by the realization that microbes drive the conversion of organic nitrogen into plant-available forms. Yet for a long time the story was told as bacteria doing the chemistry and plants merely collecting the outcome. Mycorrhizal fungi complicate that division of labor. They are not just extra straws in the soil solution; they are participants in the chemical economy that decides what becomes available in the first place.

The drama becomes most vivid when nutrients are patchy. Soils are not uniform soups. A fragment of leaf litter, a dead insect, a decaying root, a pocket of manure: each is a small island of nutrients, often rich in nitrogen but chemically guarded. Roots are comparatively clumsy at exploiting such islands. A root hair can enter a pore, but it cannot easily concentrate enzymes in a tiny location and then transport the released nutrients quickly elsewhere. Hyphae can. They can surround a patch like gauze and then act as both digestion system and pipeline.

This patch-scale behavior has been demonstrated in elegant microcosm experiments. Researchers have built simplified soil systems with discrete organic "patches" labeled with stable isotopes, then tracked where the isotopes go. In such setups, arbuscular mycorrhizal fungi—despite being obligate biotrophs dependent on plant carbon—can acquire nitrogen from decomposing organic material and move at least some of it into the plant. The fractions transferred can be modest, and the results depend on context, fungal species, and the microbial community present, but the principle matters: AM fungi are not restricted to collecting only inorganic nutrients that happen to be dissolved. They can participate, directly or indirectly, in accessing organic pools.

The mechanism is subtle, and it avoids a common misconception. AM fungi are not typically portrayed as enzyme-heavy decomposers in the way many EM fungi are. They do not, as a general rule, behave like white-rot fungi that can dismantle wood. But a fungus does not need to be the sole decomposer to be an effective miner. It can stimulate decomposition by associated bacteria and saprotrophic fungi, alter microenvironments, and capture released nitrogen quickly enough to matter. In some experimental work, AM hyphae appear to accelerate decomposition in localized patches, then siphon nitrogen from the products. If this sounds like opportunism, it is. Opportunism is often what ecology looks like up close.

The physical scale at which this plays out is easy to underestimate. A hypha may be only a few micrometers across. That thinness is not just a convenience; it is a functional advantage. Soil is a maze of pores, many smaller than a root hair. Hyphae can thread into spaces where water films cling and where dissolved ions move by diffusion.

They can maintain contact with mineral surfaces over long distances, and because they grow by tip extension, they can rapidly explore new territory when a nutrient gradient is detected. At the same time, hyphae can fuse with other hyphae of the same individual, forming networks that share resources across centimeters to meters of soil. The fungus can invest carbon in exploring a new patch and then, if it pays off, transport nutrients back to the root interface where carbon is obtained. The network makes the foraging strategy possible.

A useful mental model is to think of the hyphal network as a distributed organ that belongs, in practice, to both partners. The plant supplies the energy to build and maintain much of the fungal system (especially in AM and many EM relationships), while the fungus supplies the physical and chemical capability to reach and process nutrients. What looks like a plant "root system" in a field guide is, physiologically, a composite structure extending into soil at multiple scales: thick roots, fine roots, root hairs, and then the fungal threads that continue the gradient down into pores and mineral films.

The "hundredfold" amplification of absorptive surface area often invoked for mycorrhizal associations is a way of translating this scale into a human-friendly number. It is not that the fungus is a magic multiplier under all conditions, but that adding a web of microscopic filaments to a plant's architecture can massively increase the interface between living tissue and soil. Surface area is not a poetic abstraction; it is a currency. The number of transporter proteins, the speed of diffusion across membranes, the chance that a drifting phosphate ion will collide with a living surface before it is fixed again—these are all surface-area problems. Hyphae turn a plant's nutrient uptake from a two-dimensional boundary into a three-dimensional mesh.

Fungal mining also matters in time, not only space. Nutrient availability in soil is seasonal and episodic. A flush of nitrate can follow a rain after drought. Phosphate can become briefly more mobile when pH shifts or when microbial communities change. A dead root can release a sudden pulse of nitrogen in a localized spot. Roots respond, but slowly. Hyphae can respond quickly, extending tips toward new resources and retracting or abandoning areas that no longer pay. This flexibility has consequences for plant resilience. A plant connected to a responsive fungal network is, in effect, connected to a sensory system spread through the soil, sampling microhabitats and reallocating effort.

Forestry research has added a particularly striking piece of evidence by doing the ecological equivalent of an exclusion experiment: preventing ectomycorrhizal roots from accessing certain soil volumes and observing what happens to decomposition and nutrient cycling. In some cases, excluding EM roots can increase early-stage litter decomposition near the surface—the so-called Gadgil effect, where ectomycorrhizal fungi compete with free-living decomposers and can suppress their activity. Yet deeper in the soil profile, EM activity can be associated with oxidative processes that help mobilize nutrients from older organic matter, and when EM are excluded, enzyme activities like manganese peroxidase can drop. This is not a tidy morality tale of fungi either helping or hindering decomposition. It is a reminder that "decomposition" is not one process but many, occurring in layers, mediated by competing organisms, and shaped by who is allowed access to which carbon sources.

It also clarifies why forests can behave so differently depending on whether they are dominated by AM-associated or EM-associated trees. If a forest's dominant trees tend to partner with fungi that are adept at organic nitrogen economies, the whole system can tilt toward recycling nutrients from complex organic matter, sometimes slowing the release of nitrogen into forms that would otherwise leach away. If, instead, the dominant partnerships emphasize rapid uptake of inorganic nutrients in the soil solution, nutrient cycling can look faster and more "open," with different consequences for carbon storage and soil formation. The architecture of intimacy described earlier feeds directly into this landscape-scale behavior, because the interface dictates which fungal guilds are favored, and those guilds differ in their mining tools.

The emotional resonance of all this, oddly, is agricultural. Modern farming depends heavily on synthetic nitrogen fertilizer and mined phosphate rock. Both are finite in practical terms: nitrogen fertilizer demands energy (often fossil energy) to produce; phosphate rock is concentrated in limited geological deposits. Mycorrhizal fungi do not eliminate the need for fertilizers in intensive systems, but they alter efficiency. A plant that can extract more phosphorus from the same soil, or recapture nitrogen from organic residues more effectively, is a plant less dependent on external inputs. That matters for smallholder farmers as much as for industrial agriculture, and it matters for ecosystems asked to withstand drought, heat, and disturbed soils.

Yet the most compelling reason to pay attention is not that fungi are a sustainable technology waiting to be deployed, but that they have already been deployed for hundreds of millions of years. They are the original engineers of nutrient accessibility on land. When a fungus threads into a pore, secretes a molecule that nudges phosphate off a mineral surface, and then shuttles that phosphate back through a living pipeline to a root, it is performing a kind of chemistry and logistics that a plant alone cannot easily match. The soil is full of riches that are chemically real but biologically unavailable, and hyphae are the thin, persistent instruments that make those riches negotiable.

A handful of dark leaf fragments in the humus, a smear of iron oxide on a grain of sand, a decaying insect wing pressed into a soil aggregate: these are the kinds of things that feed forests, and they do so not because roots can swallow them, but because fungi can worry at their surfaces patiently, releasing what can be moved and storing what must wait. In that slow, localized labor—enzyme by enzyme, pore by pore—the indigestible becomes edible, and the forest's apparent solidity is revealed as a continuous act of microscopic extraction.

3.3 The Economy of Cheating

Every market has its romance: the idea that trade makes everyone richer, that specialization encourages cooperation, that a little self-interest can be harnessed into a stable peace. Underground, that romance is constantly tested. A plant can photosynthesize—turn light into sugar—but cannot easily dissolve a rock grain or pry nitrogen from a tough scrap of humus. A fungus can do astonishing chemistry and fit its hyphae into pores too small for any root hair, but it cannot capture sunlight. The trade looks inevitable. And yet inevitability is exactly what evolution does not permit. If one partner could take without paying, it would. If paying became pointless, it would stop. The most revealing fact about mycorrhizal symbiosis is not that it exists, but that it persists.

The persistence depends on two ideas that sit uncomfortably together. First, mycorrhizas are genuine mutualisms in many circumstances: plants often grow better with their fungi than without them, and fungi often reproduce better with their plants than without them. Second, mycorrhizas are not a moral category. They are relationships whose sign can flip. Under one set of conditions the fungus is a helpful supplier; under another it is a drain. Under one set of conditions the plant is a reliable patron; under another it is an unreliable client that tries to extract nutrients while starving its partner of carbon. The relationship lives on a continuum between mutualism and parasitism, and it is regulated by biological mechanisms that look, from a human distance, like incentives and sanctions.

Begin with the simplest reason cheating is tempting: the currencies are different. The plant pays largely in carbon, a versatile energy-rich resource made in leaves and transported down to roots. The fungus pays largely in mineral nutrients and water. Those currencies cannot be directly compared; there is no universal "exchange rate" between a gram of sugar and a milligram of phosphorus. The value depends on context. A plant in bright light with ample carbon but starved of phosphorus will treat phosphorus as priceless. The same plant in fertilized soil with phosphorus in abundance may find the fungal partnership a costly subscription it no longer needs. The fungus, too, experiences changing values. A fungus attached to a host that is shaded, drought-stressed, or grazed may receive little carbon and may respond by withholding nutrients, storing them, or investing in its own survival structures. Neither organism needs cognition for this. Natural selection has shaped regulatory systems so that when the balance shifts, behavior shifts.

The language of "biological markets" was borrowed deliberately to capture the idea that partners can exercise choice. A plant is not forced to allocate the same carbon to every fungal partner. A fungus is not forced to deliver the same nutrients to every root segment. Even within a single root system, there are many potential partners, and the plant can bias its investments. In experiments led by E. Toby Kiers and colleagues (2011), plants supplied with different fungal partners were able to allocate more carbon to the fungi that delivered more nutrients, and fungi could also adjust their nutrient delivery in response to carbon supply. The striking feature of such results is not that plants and fungi are selfish—that is assumed—but that their selfishness can stabilize cooperation when it is expressed as conditional investment rather than blind generosity.

Sanctions begin at the smallest scale: the interface where molecules cross membranes. In arbuscular mycorrhizae, each arbuscule is a temporary trading structure inside a root cortical cell. The plant has a lever that is almost too obvious: it can shorten the arbuscule's lifespan. Arbuscules are built, used, and then dismantled. If a particular arbuscule is not paying its way, the plant can terminate that interface and invest in new ones elsewhere. This is not a courtroom penalty; it is a local decision about where to keep an expensive membrane and transporter machinery active. In ectomycorrhizae, where exchange happens across the Hartig net and a mantle coats the root tip, the levers differ, but the principle remains: carbon flow into that tip can be adjusted, and tips can be shed and replaced as roots grow. Roots themselves are dynamic organs, and their growth patterns are part of the bargaining environment.

Sanctions also operate through the plant's control of carbon distribution. Carbon does not ooze evenly from roots like syrup from a bottle. It is delivered through phloem and allocated according to sink strength: tissues that demand carbon attract it. A fungus that stimulates more sink activity at its interface can draw more carbon, and a plant can regulate sink strength through hormonal and developmental pathways. This is one reason debates about market theory in mycorrhizas are not mere semantics. Some researchers emphasize partner choice and reward as the key stabilizers; others emphasize that carbon and nutrient flows can be governed by physiology and transport constraints, without implying anything like a "price." Florian Walder and Marcel van der Heijden (2015) argued that reciprocity is real in some contexts but not universal, and that exchange outcomes can be strongly shaped by the identities of partners and the environment. Kiers and Stuart West (2016) responded that market logic does not require perfect tit-for-tat trades in real time; it requires that, on average, more beneficial partners receive more investment. The argument matters because it keeps attention on what can be measured: who gets what, when, and under what conditions.

One reason measurement is hard is that cheating does not always look like theft. Sometimes it looks like prudence. Consider phosphorus. A fungus can take up phosphate from soil and deliver it to the plant, or it can store it in its own tissues as polyphosphate granules, a kind of internal bank account. Under reduced carbon supply from the plant, a fungus might store more and deliver less, waiting for better terms. Edith Hammer and Pål Axel Olsson (2011) found that when plant carbon availability was halved, an arbuscular mycorrhizal fungus accumulated much more phosphorus in spores and hyphae. From the plant's perspective, this looks like the fungus hoarding. From the fungus's perspective, it looks like hedging against a bad payer. The same behavior can be framed as cheating or as self-protection, depending on which side you sympathize with, which is exactly why moral language fails and economic language, used carefully, can illuminate.

The mutualism—parasitism continuum becomes clearest when soils are fertilized. In high-phosphorus agricultural soils, arbuscular mycorrhizae sometimes provide little benefit and can even reduce plant growth. This is not because the fungus has turned evil, but because the plant's cost—carbon diverted to the fungus and to maintaining the symbiotic machinery—can outweigh the benefit when phosphorus is already easy to obtain. Ecologists Nancy Johnson and James Graham, writing in the 1990s, emphasized this accounting view: the net outcome depends on developmental stage, environment, and genotype. Under low fertility, mycorrhizas can be essential; under high fertility, they can be a burden. This has practical consequences. Farmers sometimes assume that adding mycorrhizal inoculants will help crops unconditionally. In reality, a field already saturated with available phosphorus may show little response, and the fungus may behave more like a carbon sink than a nutrient supplier.

The fungus is not the only potential cheat. Plants can cheat too, and they have been remarkably inventive about it. The most dramatic form is *mycoheterotrophy*: plants that obtain carbon from fungi instead of supplying it. These plants are often pale or leafless, living in deep shade where photosynthesis is difficult. Some are orchids, which famously begin life as tiny seeds with almost no stored nutrients and depend on fungal partners to germinate. Others are entirely non-photosynthetic as adults, such as many species in the monotropes (the group that includes ghostly forest-floor plants sometimes called "Indian pipe"). Some plants are *mixotrophic*, combining photosynthesis with fungal carbon theft, a strategy common in certain orchids and forest understory plants.

How can a plant steal carbon from a fungus that is itself, in many cases, receiving carbon from trees? It does so by tapping the network indirectly. An ectomycorrhizal fungus connected to a tree receives tree carbon. If that fungus also connects to a mycoheterotrophic plant, carbon can flow into the plant. The cheating plant is, in effect, a pickpocket in a crowded bazaar: it does not mug the tree directly; it exploits the trading infrastructure. This seems like a recipe for collapse. If cheating plants proliferated, they could drain fungi, reduce nutrient delivery to trees, and unravel the mutualism.

Yet cheating plants have not taken over forests, and that fact is instructive. Network analyses and natural history observations suggest that many mycoheterotrophic plants are highly specialized. They tend to associate with particular fungal lineages rather than with whatever fungus happens to be common. That specialization can act as a constraint. If the plant's survival depends on a narrow set of fungi, it cannot spread freely into habitats where those fungi are absent or rare. Many cheaters also live in specific ecological niches—deep shade, certain soil types, particular forest compositions—where their impact on the broader network is limited. In this way, the forest tolerates a certain amount of theft without systemic collapse, not because it is forgiving, but because the cheaters are fenced in by ecological and evolutionary constraints.

Even within mutualistic plants, subtler forms of "cheating" exist. A plant might preferentially allocate carbon to roots that are already well supplied with nutrients, starving fungi elsewhere. Or it might increase root proliferation in nutrient-rich patches and reduce reliance on fungal partners. These behaviors are not deceptions; they are reallocations. But they create the same dynamic pressure: fungi must cope with fluctuating carbon supply, and plants must cope with fungi that vary in their generosity.

Spatial structure is one of the quiet regulators of fairness. If a plant's roots are colonized by multiple fungal partners that occupy distinct root regions, the plant can reward some and punish others locally. James Bever and Susan Richardson (2009) argued that such spatial segregation can help stabilize mutualism: beneficial fungi receive more carbon and thus increase in abundance, while less beneficial ones are starved. But if fungal communities are well mixed—if many fungi share the same root zones and the same carbon pool—the plant's ability to discriminate weakens. Under those conditions, less beneficial fungi can persist because they are harder to target. This is a sobering point for anyone who imagines the underground network as a perfectly managed marketplace. Management depends on separability. When everything is blended, sanctions become blunt.

There is also a deeper physiological constraint: nutrients do not always flow to the places the plant "wants" them to flow. A fungus may deliver nutrients preferentially to root segments that supply more carbon, but transport within the plant then redistributes those nutrients according to its own priorities. Likewise, carbon delivered to one root tip can, under certain conditions, be shared through fungal networks to other plant hosts. These cross-flows are part of what makes common mycorrhizal networks so fascinating and so contentious. They create the possibility that one plant could benefit from another's carbon indirectly, or that fungi could use carbon from one host to support exploration that ultimately benefits a different host. From a narrow human view, this looks like misallocation; from a fungal view, it is portfolio management; from the system's view, it is a complicated set of couplings that can either stabilize or destabilize cooperation depending on circumstances.

A concrete example brings these abstractions back to earth. Imagine a mixed forest edge where young seedlings grow under the partial shade of adult trees. The seedlings, with small leaf areas, are carbon-limited; the adult trees are relatively carbon-rich. The soil is patchy: some microsites are rich in nitrogen from decaying litter, others are poor. Ectomycorrhizal fungi connect to both adults and seedlings. If the fungi allocate more nutrients to the roots that provide more carbon, seedlings might be disadvantaged, receiving less nutrient support precisely when they need it. But fungi also benefit from keeping seedlings alive as future hosts, and the network's spatial structure may allow seedlings to tap into nutrient patches discovered by hyphae supported by adult carbon. Depending on drought, light, and soil fertility, the same network can look like a cooperative nursery system or like a competitive apparatus that amplifies inequality. The moral interpretation flips; the biology does not.

What keeps the system from devolving into pure exploitation is not kindness but feedback. Carbon is not given unconditionally. Nutrients are not given unconditionally. When costs exceed benefits, partners adjust. Sometimes they do so by withdrawing: the plant reduces colonization, the fungus shifts to survival structures, the interface is dismantled. Sometimes they do so by bargaining harder: allocating resources preferentially, changing the local chemistry of the interface, altering root architecture, or investing in different parts of the network. The elegance of mycorrhizal symbiosis is that these feedbacks are embedded in anatomy and physiology. The arbuscule is built to be temporary, the Hartig net is built to be intercellular, hyphae are built to explore and retract, and the currencies are moved by transporters whose expression can be dialed up or down.

A forest floor in late summer can look serene: ferns unrolling, needles softening underfoot, a few mushrooms rising after rain. Under that calm is a constant negotiation over molecules too small to see. Carbon is offered, withheld, redirected. Phosphorus is delivered, banked, delayed. Nitrogen is extracted from stubborn organic matter and routed along living threads. Fair trade is not a stable state but a moving target, enforced locally by countless tiny decisions made through membranes, enzymes, and growth patterns—the only kind of enforcement evolution ever really trusts.

Chapter 4

Maps of the Invisible

To understand the forest, we must first abstract it, stripping away the bark and dirt to reveal the mathematical skeleton beneath. By translating biological tangle into topological structure, we discover that the underground world is governed by the same strict laws of connectivity that rule the internet, airline routes, and the human brain.

4.1 From Soil to Syntax

A forest floor is easy to misunderstand because it looks still. Leaves fall, needles soften into duff, a few mushrooms appear after rain and collapse into jelly, and the rest seems like quiet matter waiting its turn. Yet much of the forest's work happens in the thin zone where roots press into soil and fungi press back: a crowded interface of living filaments, dying cells, mineral grains, water films, air pockets, and bacteria. If the forest were only what we can see, it would be a sparse arrangement of trunks separated by metres of empty space. Underground, it is closer to a city at rush hour: densely occupied, intensely negotiated, and full of routes.

The difficulty is not merely that the routes are hidden. It is that they are too numerous, too irregular, and too entangled to hold in the mind as a picture. A single root tip can be sheathed in fungal tissue; that fungus can branch into thousands of threads; those threads can fuse, split, or thicken into cords; and along the way they touch other roots, other fungi, and the microscopic architecture of the soil itself. To speak clearly about such a system, the first step is a kind of translation: turning messy biology into a small set of terms that can be counted, compared, and argued about without constant hand-waving.

That translation has a name that sounds more abstract than it is: a *graph*. In everyday English a graph is often a plot, a curve, a line climbing or falling. In mathematics, a graph is a different thing: a collection of *vertices* (also called *nodes*) connected by *edges* (also called *links*). Once you accept that definition, the underground begins to look less like a tangle and more like a structured object you can interrogate. Which trees are connected at all? Which are connected by many routes, and which depend on a single fragile corridor? If something is cut—by drought, disturbance, or disease—what remains connected, and what becomes isolated?

There is an older story behind this vocabulary, one that begins above ground in a river city. In 1736 Leonhard Euler published a solution to a puzzle about Königsberg (now Kaliningrad), where the Pregel River divided the town into landmasses linked by seven bridges. The question was playful: could a person take a walk that crosses every bridge exactly once and returns home? Euler's answer mattered less than his method. He stripped away street names, distances, and the picturesque detail of the riverbanks, keeping only the pattern of connection: landmasses became points, bridges became lines. By proving that the walk was impossible from the structure alone, he showed that some questions are not about *where* things are, but about *how* they are connected.

That same stripping-away is what makes the underground discussable. A forest is not a river town; a mycorrhizal network is not a set of bridges. But the move is similar: replace the overwhelming continuity of soil with a symbolic skeleton in which connection is explicit. The point is not to deny biology. It is to choose a language in which biology can be tested against alternatives, rather than admired as complexity.

Start with the most basic choice: what counts as a node? In a simple forest graph, each tree can be treated as a node. The links between them represent some meaningful connection mediated by fungi: shared fungal partners, direct hyphal continuity, or evidence of transfer. If the question is about whether seedlings have access to an established network, the tree-as-node picture is natural. If the question is instead about the fungus—which genetic individual is doing the connecting, which mycelium spreads where—then fungi themselves can be nodes, and the trees become the resources they connect. Both are legitimate, and they do not always lead to the same conclusions. The graph is not the underground; it is a particular way of asking it to speak.

Edges, too, require decisions. An edge can be *binary*: either two nodes are connected or they are not. This is the most austere model, the one Euler would recognise. But the forest is rarely so simple. Some links are thick cords capable of moving water and solutes efficiently; others are thin hyphae that may be transient. Some connections are direct physical continuity; others are ecological co-occurrence that might not imply active exchange. To keep that nuance without abandoning the clarity of a graph, network science allows *weights*. A weighted edge carries a number: thickness, length, estimated conductance, probability of connection, frequency of shared fungal genotypes, or any other measurable quantity that matters to the question at hand. In a weighted graph, not all links are equal, and that inequality can be the story.

Once nodes and edges are named, the next terms describe local structure. The *degree* of a node is simply the number of edges touching it: how many neighbours it has. Degree is a modest concept with dramatic consequences. A sapling connected to one fungal partner has degree one; a veteran tree connected to many partners and many neighbours has high degree. Degree is tempting to interpret as importance, but it is best treated first as geometry: it tells you how many immediate options a node has for interaction, whether that interaction is carbon transfer, nutrient access, or pathogen exposure.

Degree also invites a question that has become a kind of refrain in network science: what does the *degree distribution* look like? If most nodes have about the same degree, the network is comparatively uniform. If a few nodes have very high degree while most have low degree, the network has *hubs*. Networks with hubs are often called *scale-free* in a technical sense: there is no single characteristic degree that typifies the whole system. This matters because hub-dominated networks behave differently under stress. Random failures tend to hit low-degree nodes simply because there are more of them; targeted losses that remove hubs can fracture the network quickly. In the forest, this becomes a way to phrase a worry that is otherwise hard to quantify: whether losing a handful of highly connected old individuals could change connectivity in ways that losing many small individuals would not.

To move beyond immediate neighbours, graphs use the idea of a *path*. A path is a sequence of edges that leads from one node to another. In a forest, a path might represent a potential route along which resources could move, information could propagate (in the modest sense of chemical signalling, not intention), or infection could spread. Path length is counted in edges (how many steps), not metres. That sounds odd until you notice what it captures: the number of intermediary nodes involved. A tree connected to another by one fungal bridge has a path length of one; a tree connected only through a chain of three other trees has a path length of four. This is the seed of a famous observation about many real networks: despite their size, typical path lengths can be surprisingly short, a property associated with *small-world* networks. Whether any particular forest network truly has small-world structure is an empirical question; the useful point here is that path length gives a way to state what "close" means without pretending the soil is transparent.

Paths can loop back on themselves, forming *cycles*. A cycle is what it sounds like: you can start at a node, follow edges, and return to the start without retracing your steps. Cycles are redundancy made visible. In a strictly tree-like network (in the graph-theory sense), there is exactly one path between any two nodes, and there are no cycles. In a network with cycles, there are alternative routes. Alternative routes can be costly to build and maintain in living tissue, but they can also be a form of insurance. If a single corridor is broken, the connection might still exist through a different route.

That leads to another term that sounds technical but captures a basic intuition: *connectivity*. A graph is connected if every node can reach every other by some path. If not, it breaks into *components*: groups within which everything is mutually reachable, but between which there is no route. A forest with multiple components is not necessarily unhealthy; it might simply be a set of neighbourhoods that do not share fungal partners. But components become crucial when thinking about disturbance. A firebreak, a road cut, soil compaction, or the death of a few key trees can turn one connected system into several smaller ones, changing who can share with whom and who is stranded.

Connectivity can be sharpened further by asking which nodes or edges are structural linchpins. In graph theory, an edge whose removal increases the number of components is sometimes called a *bridge*. A node whose removal does the same is a *cut-vertex*. The terms are blunt, almost anatomical: remove this, and the body falls into separate parts. In the underground, identifying bridges and cut-vertices is a way of turning a general claim ("some connections are critical") into a testable one ("this connection is critical under this definition of edge").

Already, however, an uneasiness creeps in: are we allowed to treat the underground as a set of discrete points joined by lines? Hyphae are continuous, soil is granular, roots branch fractally, and much of the action involves gradients rather than switches. The answer is that the graph is not a rival description to biology. It is a compression, like a subway map. A subway map distorts distance and geography, but it preserves something we often need more: which stations connect, where you must change lines, and how many steps it takes to get from here to there. The underground has its own geography, and sometimes the geography matters deeply. Yet many questions begin with the simpler matter of reachability: can this tree access that fungal pathway at all?

Even the simplest subway map, though, hides variation in track capacity. That is why weighted edges matter, and why weights are not merely decorative numbers. In fungal systems, thickness can stand in for conductance, since wider cords can carry more flow. Length matters because long links may be costly, slow, or vulnerable. Some researchers distinguish between *structural* weights (properties of the physical cord) and *functional* weights (estimated traffic along it), since a thick road is not the same as a busy road. The distinction matters for interpretation. A cord that is anatomically impressive might not be doing much today; a thin cord might be crucial during a pulse of nutrient movement.

There is a further choice that shapes the graph before any measurement begins: whether links are *directed* or *undirected*. An undirected edge simply means two nodes are connected; a directed edge means there is a direction to the relationship. In rivers, direction is obvious. In ecology, it is often conditional. Carbon typically flows from photosynthesising plants into fungi, but transfer between plants via fungi can be bidirectional and context-dependent. Nutrients such as nitrogen and phosphorus can move from fungal foraging zones into roots, but the balance of exchange shifts with season, light, and stress. When the direction is uncertain or variable, an undirected edge can be the honest representation. When experiments show net transfer in one direction under specified conditions, a directed edge becomes a way to capture that asymmetry without claiming it is permanent.

One concrete example of how "syntax" meets soil comes from the use of isotopic tracers to follow carbon in the field. In the late twentieth century, Suzanne Simard and colleagues used reciprocal isotope labelling to examine carbon transfer between ectomycorrhizal tree species. The details are technical, but the conceptual result is simple enough for a graph: there exist circumstances in which carbon fixed by one plant can later be detected in another, and the pattern can be consistent with a hyphal pathway rather than mere diffusion through soil. Once that possibility is established, the network vocabulary stops being metaphor and becomes a scaffold for questions. If transfer is possible, along which paths is it likely? Which nodes have multiple routes, and which have only one? Does the network favour certain distances or neighbourhoods? Does the structure change with age, species composition, or disturbance?

Of course, the forest does not hand over its edges neatly labelled. Much of the labour in this field lies in deciding what the graph should be faithful to. A physical connection is one kind of edge; a shared fungal genotype is another; a demonstrated transfer under controlled conditions is a third. Conflating them makes for vivid slogans and confused arguments. Keeping them distinct makes for slower prose and better science.

The payoff of this careful vocabulary is that it lets us be precise about claims that otherwise drift into myth. Consider a sentence like “the forest is all connected.” In graph terms it becomes a set of separable propositions. Is the forest connected under the definition “trees are nodes, and an edge exists if two trees share at least one fungal partner”? Is it connected under the stricter definition “an edge exists only if a continuous mycelial individual physically links both roots”? Is it connected when edges are weighted by estimated conductance and links below a threshold are treated as absent? Each definition describes a real aspect of biology, and each yields a different answer. The point is not to choose the most dramatic one. The point is to say which one is being used.

The same clarity helps with a different, more human temptation: to treat networks as intentional coordinators. A network, in everyday speech, suggests planning: a network of friends, a network of collaborators, a network of spies. In the soil, network structure can arise from local growth rules, resource gradients, and physical constraints without any central control. Graph language helps here because it is descriptive before it is interpretive. A node with high degree is not a “leader”; it is a node with many links. A hub is not a “decision-maker”; it is a concentration of connectivity. The forest may still behave in coordinated ways, in the sense that local interactions can produce system-level patterns, but the words themselves do not smuggle intention into the story.

There is also a subtle ethical comfort in the vocabulary. Forest management, conservation, and restoration are full of arguments about what matters: species, ages, densities, patch sizes, corridors. Network terms add another dimension without replacing the old ones. A stand can be diverse and still fragile if connectivity rests on a few bridges. A stand can be uniform and still resilient if it contains many cycles. These are not moral properties; they are structural ones. But structure often determines how damage propagates, and how recovery proceeds.

A final term completes the basic toolkit: the difference between *topology* and *geometry*. Topology is about connection pattern regardless of physical distance; geometry cares about metres, angles, and spatial embedding. A graph captures topology. The underground lives in geometry. Neither view is sufficient alone. A short topological path could still be a long physical route if edges are long. A highly connected node could be spatially central, or simply genetically promiscuous with fungal partners. The art is to let topology ask crisp questions while geometry keeps the answers honest.

With these few words—node, edge, weight, degree, path, cycle, component, bridge, direction—a system that once looked like pure tangle begins to admit to structure. The forest floor remains opaque, but it is no longer mute.

From roots and hyphae to nodes and edges

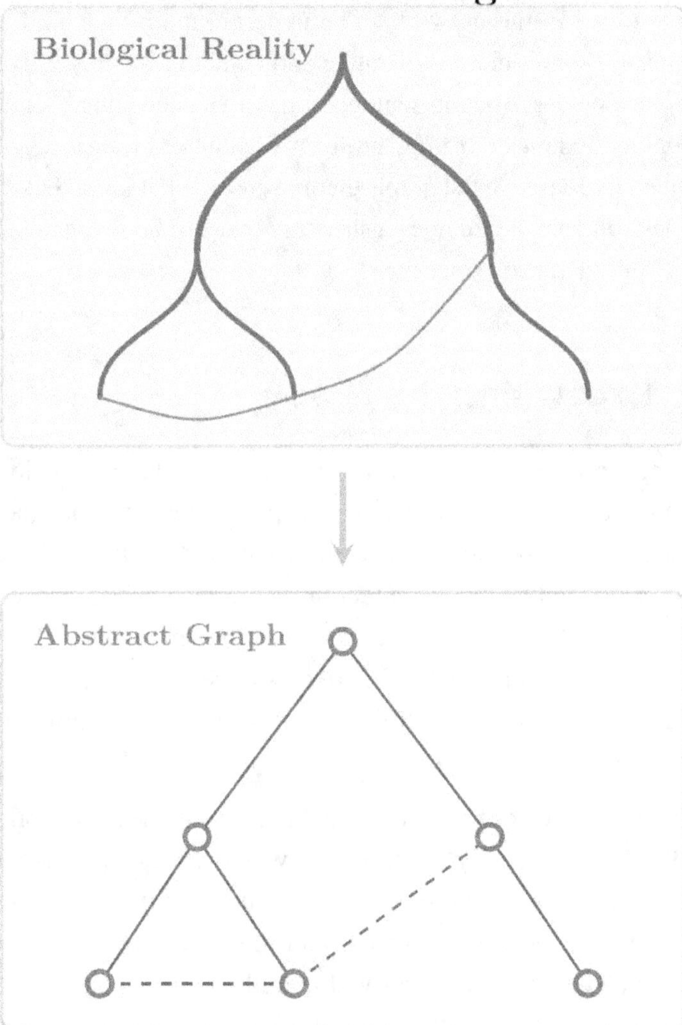

The trick, once learned, is hard to unsee. A cluster of trees becomes a set of nodes with different degrees. A patch of soil that dries first becomes a stressor that may sever bridges. A thick cord becomes not only a strand of tissue but a high-capacity edge whose loss would change travel in the network. The underground remains alive, opportunistic, sometimes cooperative and sometimes competitive, but it can now be described in sentences that invite checking: connected or not, redundant or brittle, uniform or hubbed, rich in cycles or arranged like a tree in the graph-theory sense. What looked like mere darkness underfoot becomes a structure that can be mapped, argued about, and, with care, measured.

4.2 Excavating the Graph

Digging up a network is a little like trying to learn a melody by smashing the piano. The moment a spade bites into soil, the very filaments one hopes to follow tear, smear, and vanish into crumbs. Roots snap; hyphae that were taut bridges a second ago slacken into invisible lint. Even careful excavation changes moisture and oxygen, and those changes prompt living tissues to behave differently. If the goal is to map connections, the usual act of looking becomes a form of vandalism.

That is why much of the work of making subterranean networks legible has been, strangely, an argument with darkness. Soil is opaque not simply because it is brown, but because it is full of boundaries: mineral grains, organic fragments, and air spaces scatter light so thoroughly that the view collapses within millimetres. The underground is also crowded with look-alikes. A fungal hypha and a plant root hair can occupy the same scale. A dark cord can be a fungal rhizomorph, a rootlet, or a decomposing fibre. Bacteria cling to everything. If you peel a slice of soil and stain it, you may see threads everywhere and still not know which threads matter.

So the problem of "excavating the graph" has two faces. One is physical: how to observe a fragile, three-dimensional object without destroying it. The other is conceptual: how to decide, from partial evidence, what counts as a meaningful connection. The second difficulty often arrives disguised as a technical detail, but it is the place where maps can become myths. A line drawn between two trees can mean "the same fungus has been detected on both," or "a continuous mycelium has been traced," or "a net transfer of carbon occurred under these conditions." Those are not synonyms. They are different edges in different graphs.

The earliest attempts to understand mycorrhizae were necessarily intimate, even artisanal. In the nineteenth century, when microscopes improved and staining became a craft, botanists began to notice that many roots in healthy soils were not naked at all. They were wrapped, invaded, or threaded with fungal tissue. Albert Bernhard Frank, working in the 1880s, did something that still feels modern: he insisted that these fungi were not mere disease but a partnership, and he gave the relationship a name, *mycorrhiza*—literally "fungus-root." Frank could not map a network in the contemporary sense, but he established a crucial point: if roots and fungi are bound together as a normal state of life, then the basic unit of underground ecology is not an isolated tree.

For a long time, the best one could do was to infer networks from fragments: a root tip colonised by this fungus here; another colonised by the same-looking fungus there. But "same-looking" is a treacherous category. Many fungal species produce similar morphologies on roots, and even within a species, appearance can shift with conditions. Mapping required a way to identify who is who without relying on the human eye's confidence.

Molecular tools changed the landscape. When DNA sequencing became accessible enough to use routinely in ecology, the question "which fungus is on this root tip?" could be answered by genetic markers rather than by shape. Researchers could sample roots from different trees and ask whether the same fungal genotype appeared on more than one host. That is a powerful clue to connection, because a genotype shared across hosts suggests a single fungal individual, or at least a close kinship capable of bridging. It also makes it possible to build a different kind of network, one that is explicitly bipartite: one set of nodes is plants, the other is fungi, and edges connect plants to the fungi that colonise their roots. This plant—fungus graph can then be "projected" into a plant-only network (two plants are linked if they share a fungal partner) or a fungus-only network (two fungi are linked if they share a plant host).

That projection step is not a mathematical nicety; it can change the story dramatically. Imagine an oak woodland where many trees host a few common fungi, while rare fungi specialise on particular microhabitats. In the plant projection, most trees share those common fungi and appear richly interconnected, with a fairly even number of links per tree. In the fungus projection, those common fungi become hubs, linked to many plant nodes, and the distribution of fungal degrees can look heavy-tailed: a few fungi with many links, many fungi with few. Both statements can be true at once, because they refer to different graphs extracted from the same biology. The network you "see" depends on whether you are asking, in effect, "how socially connected are the trees?" or "how widely distributed is each fungus?"

This dependence on definition is not a weakness. It is a warning label. The underground is not obliged to offer a single, definitive wiring diagram. It offers a set of relationships, and the mapper must declare which relationship is being mapped.

Even with genetic identification, there remains a stubborn gap between "shared partner" and "physical continuity." A fungus can have spores and fragments dispersed through soil; the same species can appear on multiple roots without those roots being linked by a continuous mycelium at that moment. Genotyping can help narrow the question from species to individual, but even then, sampling is sparse. A fungal individual can extend metres, but samples are points. The map is a reconstruction, not a photograph.

Physical imaging tries to close that gap by treating the mycelium as something that can be seen, not merely inferred. There are two broad approaches. One is to bring the network into view by extracting it: soil cores, root systems, and hyphal mats removed and processed. The other is to look through soil without opening it, using methods borrowed from medical imaging and materials science.

Extraction sounds straightforward until you try it. Soil is not a coherent block but a fragile architecture of aggregates, whose stability depends on moisture and organic glues. Fungal hyphae are on the scale of micrometres, and they can be torn by the slightest shear. A single core sample can contain a bewildering density of biological threads, and each thread can branch, fuse, or run alongside another. To turn such a sample into a network, one must first separate signal from background.

Here, the challenges become unexpectedly similar to those faced in fields like neuroscience, where researchers try to reconstruct neurons from microscopy images. A fungal network is, in a literal sense, a curvilinear structure: long, thin lines that cross, branch, and sometimes merge. Automated image analysis can help, but only if the pipeline respects the peculiarities of fungal tissue: cords that vary in diameter by orders of magnitude, faint hyphae that disappear into noise, and overlapping structures that confuse segmentation.

A typical computational pipeline begins by enhancing line-like features in the image, accentuating filaments while suppressing blob-like noise. Then it segments the enhanced image into foreground and background, often with algorithms that treat the image like a landscape of intensities and "flood" it into regions. After segmentation, the filaments are reduced to a skeleton: a one-pixel-wide representation that preserves connectivity while discarding thickness. Branch points become candidate nodes; the segments between them become candidate edges. Finally, the skeleton is pruned and cleaned to remove spurious twigs created by noise.

This process is both miraculous and dangerous. Miraculous because it can convert an impossibly dense image into something countable, even on the scale of hundreds of thousands of edges. Dangerous because each step introduces assumptions: what counts as filament, what threshold separates it from background, when two nearby threads are treated as crossing versus merging. A human eye can often resolve such ambiguities by context; an algorithm must be told what "context" means.

One way to keep the graph tied to biology is to make it weighted. If thickness can be estimated from the original image (for instance by measuring how wide the filament appears before skeletonisation), then the resulting graph can carry an approximation of conductance or transport capacity. Edge length is also meaningful: long edges may correspond to long cords, and their length is directly measurable in the skeleton. With weights, the graph is no longer just a yes-or-no connectivity map; it becomes a transport network whose edges have costs and capacities.

Even then, there is a further conceptual fork: a weighted graph can represent structure or function. A structural network uses weights derived from anatomy—thickness, estimated conductance, perhaps wall properties. A functional network uses weights derived from predicted or observed flows—nutrient traffic, water movement, or carbon allocation. The same topology can tell different stories under these two weightings. A thick cord may be capable of high flow but currently idle; a thin corridor may be carrying a disproportionate share under stress. For a fungus responding to a patchy food supply, function can change faster than structure.

If extraction and microscopy have a flavour of dissection, non-destructive imaging has a flavour of espionage. X-ray computed tomography, for example, can produce three-dimensional reconstructions of soil structure by measuring how much X-ray is absorbed as it passes through the sample from many angles. The technique is excellent at resolving mineral grains, pores, and sometimes thicker roots. Fungal hyphae, however, are often too thin and too similar in density to the surrounding organic matter to stand out clearly. Contrast agents can help, but they introduce their own distortions. Magnetic resonance imaging can sometimes visualise water distribution, which is relevant because water films outline the spaces in which hyphae live and through which they transport. Yet again, one ends up mapping proxies: habitats for networks rather than networks themselves.

These methods are improving, and their greatest value may be in combining with other data rather than replacing it. A three-dimensional pore map can tell you where hyphae *could* have grown, and where they would have been forced to detour. A genetic survey can tell you which fungi *were* present. A physiological tracer experiment can tell you where resources *moved*. The graph that emerges is then a triangulation: not a single instrument's verdict, but an agreement among imperfect witnesses.

A concrete example of how these witnesses complement one another comes from field isotope labelling experiments that track carbon transfer. When one plant is supplied with a carbon isotope label, the appearance of that label in a neighbouring plant can suggest transfer through a shared fungal connection. But alternative routes exist: root exudates into soil, microbial intermediaries, or physical leakage. Designing experiments that discriminate among routes requires careful controls, including plants that cannot form the relevant mycorrhizal associations. In some of the influential field work in this area, such controls have been used to argue that a direct hyphal pathway is the most plausible explanation for observed transfers under the tested conditions. The graph language is useful here because it provides a disciplined way to phrase what is being claimed: not that "trees share carbon" in general, but that under specified circumstances, a path exists in the underground network capable of moving carbon between hosts.

Yet even the best tracer experiment does not hand you a full map. It provides evidence that at least one path exists between two nodes. The network could be sparse or dense; it could rely on one fungus or many; it could contain cycles or be mostly tree-like. Tracer evidence is like learning that two cities have a direct flight. It does not tell you the entire airline network.

The most revealing maps tend to come from studies that accept this incompleteness and treat mapping as a statistical problem. Rather than attempting to expose every hypha, researchers sample many root tips, many soil points, many trees across a plot, and they reconstruct a probabilistic network: links with confidence levels, weighted by frequency of co-occurrence, genetic identity, or observed transfer. Such maps can be compared across time, disturbance, or management regimes. They can also be interrogated with the basic questions of network science: does the network fragment into components under drought? Do certain fungi act as hubs in the fungus projection? Does connectivity correlate with seedling survival or with the spread of pathogens?

A subtle technical detail becomes emotionally charged here: sampling bias. Old trees have larger root systems and can be easier to sample. Some fungi colonise root tips in ways that are more visible or more readily amplified by standard genetic primers. Soil heterogeneity means that a core taken a few centimetres away can contain a different community. If sampling is uneven, hubs can be manufactured by effort rather than by biology. The map begins to resemble the researcher's footsteps.

Acknowledging bias does not paralyse the project; it makes it honest. Network science has long wrestled with the fact that many large networks are observed incompletely. The internet is mapped by probing routes that can change; social networks are inferred from partial records; ecological networks are assembled from sightings that reflect observer attention. Techniques exist to estimate how robust conclusions are to missing edges and nodes. In forests, one can ask: if a certain fraction of edges were missed, would the identification of a hub remain? If the threshold for considering a link "real" is tightened, does the network remain connected or break into components? These are not purely mathematical games. They are a way of keeping the romance of the "wood-wide web" tethered to what has actually been observed.

There is, finally, a physical limit that has a philosophical sting: some networks are not stable objects. Fungal mycelia grow, retract, fuse, and die. Roots proliferate and are shed. Connections can be seasonal, or even daily, responding to moisture pulses and temperature. A map drawn from today's data can be accurate and still be obsolete. The underground is not a wiring diagram but a shifting compromise between opportunity and constraint.

So what is a map for, if not permanence? A good map does not pretend to freeze the forest. It captures regularities: tendencies of connection, recurring hubs, typical path lengths, the existence of bridges that are often present even if their precise strands change. It also captures the range of possible configurations under different conditions. The power lies not in pinning a living network to the page, but in turning the invisible into something that can be compared: one plot against another, one season against the next, one forest before and after disturbance.

Somewhere in that comparative work, the forest begins to acquire a new kind of visibility. Not the visibility of a root held in the hand, but the visibility of a pattern that survives imperfect sampling and shifting tissues: clusters that persist, corridors that repeatedly carry the traffic of resources, and particular associations that keep reappearing as if the soil itself were offering a preference. The graph, excavated with genetics, imaging, and careful inference, becomes less like a drawing of threads and more like a portrait of constraints under which life has learned to connect.

4.3 The Geometry of Giants

Walk into an old forest and the inequality is obvious. A few trees hold the canopy like pillars; a scattering of veterans carry lightning scars and old fire wounds; saplings crowd the gaps. Above ground, size announces itself. Underground, size becomes something less visible and, in some ways, more consequential: a veteran's roots occupy more space, touch more microsites, and offer more docking points for fungal partners. If a network is made of nodes and links, then age and girth have a geometric meaning. Some nodes become places where many paths can pass, not because they intend to coordinate anything, but because they are simply hard to avoid.

Network science has a name for this kind of disproportion: *centrality*. Centrality is not a single measure but a family of ways to formalise what it means for a node to be "important" in the structure of a network. The underground invites this formality because it resists ordinary intuition. A tree can look isolated from its nearest neighbour above ground and still sit at the centre of a dense web of shared fungi. Another can stand amid a thicket and yet be relatively peripheral if it hosts few compatible partners. Centrality turns these possibilities into quantities that can be compared and, crucially, argued about.

The simplest centrality measure is *degree centrality*: count the number of links touching a node. In a tree-focused network, degree centrality answers a blunt question: how many other trees is this tree directly connected to, under the chosen definition of "connected"? In a fungus-focused network, degree centrality asks how many host plants a fungal individual touches, or how many other fungal individuals it co-occurs with through shared hosts. Degree is the first handle because it is local and concrete: one can often estimate it from sampling with fewer assumptions than more elaborate measures.

Yet degree has a moral hazard. It is easy to hear "high degree" as "powerful" and "low degree" as "unimportant." The safer interpretation is structural: high-degree nodes are potential *junctions*. They create options. They also create dependencies. In a forest network with hubs, many short paths pass through the same few nodes simply because those nodes offer the easiest routing. That is where centrality becomes less like social prestige and more like plumbing.

To see why, imagine two ways of wiring the same number of nodes. In one, connections are fairly uniform: each tree has, say, three fungal links to neighbours, distributed evenly. In the other, most trees have one or two links, while a few have ten or twenty. The second network has hubs. The difference is not cosmetic. In the uniform network, removing one node tends to have modest effects; alternative routes often exist. In the hubbed network, losing a hub can sever many paths at once. A hub is a concentration of connectivity, and concentrations are always points of vulnerability.

This relationship between hubs and vulnerability is often framed through the idea of *robustness*. A robust network keeps most nodes mutually reachable even when some nodes or edges are removed. Robustness depends on the type of loss. Hubbed networks can be remarkably robust to random losses, because random losses usually hit low-degree nodes. But they can be fragile to targeted losses that pick out hubs. In human systems, this is the difference between a few random street closures and the collapse of a major bridge. In forests, targeted loss can happen when disturbances preferentially remove the largest trees: logging, storms that topple tall stems, droughts that kill trees at the edge of their hydraulic limits, bark beetles that select certain size classes. A network with hubs does not need intention to be "coordinated"; it only needs geometry that concentrates routes.

Degree centrality, though, does not tell you everything about routes. A node can have a modest degree and still be a critical passage if it lies on the only connection between two otherwise dense clusters. Graph theory describes this with the idea of a *cut-vertex*: remove it, and the network splits into separate components. The parallel concept for edges is a *bridge*. These are the places where redundancy is absent. In the underground, bridges can be literal: a narrow corridor of suitable soil moisture; a strip of intact organic layer between compacted patches; a single compatible fungal individual spanning two neighbourhoods of trees. They can also be methodological artefacts if the map is incomplete, which is why claims about critical nodes must be held with care. But the concept itself is invaluable, because it captures a kind of importance that degree misses: the importance of being a chokepoint.

A more global way to quantify chokepoints is *betweenness centrality*. Betweenness counts, for each node, how often it lies on shortest paths between other nodes. "Shortest" here is shortest in the graph, not necessarily in metres: the path with the fewest edges, or the lowest total weight if weights represent costs. A node with high betweenness is not necessarily the one with the most neighbours; it is the one through which many efficient routes pass. In an underground transport network, high betweenness can indicate a conduit through which resources *could* flow efficiently, or through which pathogens *could* spread efficiently, or through which the network's connectivity is channelled.

Betweenness invites a more realistic picture of veteran trees. An old tree is not just larger; it is also more likely to be an *intermediary* because its root system overlaps with multiple cohorts and microhabitats. It can sit at the boundary between a moist hollow and a drier ridge, between shade-tolerant understory and sun-demanding edge, between soil types that host different fungal communities. In such a position, the veteran becomes a route not because it is a benefactor, but because it is a geographic and biological intersection.

There is another centrality measure that suits the ecology of exchange: *closeness centrality*. Closeness asks how near a node is, on average, to all other nodes in the network, measured by path length. A tree with high closeness can reach others through relatively few steps. In a network where edges represent potential exchange, high closeness suggests that the tree is well-placed to interact broadly, to access diverse fungal foraging zones, or to be exposed to diverse signals and microbes. In a disturbance, it may also be well-placed to be infected quickly. Closeness is a reminder that "central" can mean "connected" in ways that are not necessarily comforting.

Even these measures are still, in a sense, topological: they treat edges as abstract connections. The underground adds a further layer because connections are not only present or absent; they carry flows. That is where weights and the idea of a transport network become essential. A thick fungal cord is not the same as a thin hypha. A short connection is not the same as a long one. If weights represent estimated conductance, then a "shortest path" might mean a path of least resistance rather than fewest steps. Centrality computed on such a weighted network begins to resemble physiology: which nodes lie on high-capacity routes, and which are served only by narrow capillaries?

At this point, the metaphor of "mother trees" can be handled more carefully. The phrase is emotionally appealing, and it has helped bring public attention to the reality that forests are not merely collections of individuals. But emotionally appealing language can blur categories. A "mother" suggests intention and care. Centrality provides a different kind of claim: that some older trees, by virtue of their position and connectivity, can act as hubs or chokepoints. That claim is both more limited and more testable. A hub can facilitate transfer without deciding to. A chokepoint can be vital without generosity.

A concrete example of "giant geometry" comes, unexpectedly, from fungi rather than trees. In northeast Oregon, work on *Armillaria ostoyae* documented an individual fungal genet sprawling across roughly 965 hectares, with isolates separated by kilometres. Popular accounts dubbed it the "Humongous Fungus," and the nickname stuck because it forces the imagination to stretch: one organism on the scale of a landscape. For network thinking, the interest lies not only in size but in structure. A giant genet implies long-distance connectivity: a component that remains connected over large spatial scales. It also raises questions about redundancy. Is the genet a loopy mesh with many cycles, or a sparse set of corridors vulnerable to breaks? How does it maintain transport across such distances? Which parts are reinforced, and which are thin exploratory threads?

The case is instructive because it separates two notions that are often conflated: *being large* and *being central*. A giant fungal individual can be large but not necessarily central in a plant-focused network if it connects mainly to a limited set of hosts. Conversely, a fungal individual could be relatively small yet central if it bridges many host neighbourhoods. Size and centrality are different axes. Giants can be central, but centrality has to be measured, not assumed.

Transport physics gives a further reason why hubs and backbones can emerge. In cord-forming fungi, growth is not merely adding new tips at the margins; it also rearranges internal flows. Because fluids in hyphae are effectively incompressible, adding volume in one place and taking up water in another induces currents through the network. Models of fungal transport have explored how these growth-induced flows can be inferred and how they might shape the network's architecture: cords predicted to carry larger currents tend to be reinforced, thickening into stable highways while less-used paths remain thin or regress. The result is a kind of self-organised hierarchy: a backbone of high-capacity links with a halo of exploratory filaments. No overseer is required. The network writes its own infrastructure in response to where resources are, where growth is happening, and which routes minimise energetic cost.

This is where loops become more than decorative complexity. A loopy network can be more expensive to build than a tree-like network because it contains extra material. But loops can improve mixing, provide alternative routes when a corridor is cut, and allow traffic to be rerouted around blockages. Optimisation theory for transport networks formalises this trade-off. Under strict constraints, the most efficient network can become path-like, almost like a branching tree: minimal redundancy, minimal cost. Relax the constraints or add a premium on mixing and robustness, and loops can become optimal despite their cost. In a forest, the constraints change with moisture, temperature, and damage. What looks like "wasteful redundancy" in one season can become survival infrastructure in another.

Centrality becomes more informative when combined with this idea of trade-offs. A node that is central in the structural network (high-capacity edges) might not be central in the functional network (high-traffic edges) at a particular moment. During drought, for instance, some pathways may close as soils dry, shifting traffic to deeper, moister routes. In that situation, the "central" node is not necessarily the one with the thickest cords in an anatomical map, but the one that remains connected through the surviving wet corridors. Functional centrality can change faster than structural centrality, and that difference matters for interpreting what maps mean.

This is also where the human stakes become clear. Forestry and conservation often focus on what is visible: species composition, basal area, canopy cover, the number of old trees per hectare. If veteran trees act as high-centrality nodes in underground networks, then removing them is not merely subtracting biomass; it can rewire connectivity. Rewiring can affect regeneration if seedlings lose access to established fungal partners. It can affect resilience if the network loses cycles and becomes more tree-like in the graph-theory sense: efficient under stable conditions, brittle under shocks. It can also change disease dynamics, potentially reducing some transmission routes while opening others through stressed and damaged tissues.

None of this demands a story in which the forest is a single harmonious superorganism. Networks can coordinate without benevolence. They can also coordinate harm. A highly connected pathway that helps seedlings obtain nutrients can also help pathogens move. A hub that stabilises exchange can also become a point where stress concentrates. Centrality is a structural fact; its ecological meaning depends on what is moving along the edges.

One of the most revealing ways to think about a veteran tree, then, is not as a wise elder but as a piece of infrastructure that has been accruing for decades. Every year of growth adds root surface, adds mycorrhizal opportunities, and adds overlap with other roots. Every disturbance survived adds scars that alter microhabitats and microbial communities. The tree becomes a thick knot in the forest's below-ground fabric. Knots are good at holding things together, but they are also places where tension gathers.

A forest that contains giants has, in a literal mathematical sense, a different geometry from a forest without them. The difference can be measured as distributions, centralities, and vulnerabilities, and it can also be felt as a change in how the underground responds to stress: whether it fragments into isolated components or keeps a connected backbone; whether nutrients can be rerouted when a corridor fails; whether seedlings inherit a neighbourhood rich in partners or must begin in isolation. The veteran's influence is not a mystical force travelling through soil. It is the ordinary consequence of accumulated connections, written in the quiet arithmetic of nodes and edges, and expressed in the survival of living things that happen to be near it.

Chapter 5

Rivers of Sugar and Signal

Beneath the static appearance of a forest lies a frantic infrastructure of transport where resources follow physical gradients rather than biological altruism. This chapter traces the movement of vital fluids and chemical messages through fungal pipelines, evaluating how trees physically subsidize their offspring and potentially alert neighbors to impending doom.

5.1 The Economy of Shadow and Light

Walk into an old forest at noon and you meet two climates at once. Up in the canopy, needles and leaves are flooded with light; air moves; surfaces warm; sugars are made in abundance. A few metres below, the understory sits in a green dusk. There, a seedling may be perfectly healthy in every visible way—straight stem, intact buds, no sign of browsing—and yet live on an energy ration so tight that one cloudy week can matter.

The basic arithmetic is simple. Photosynthesis is the process by which plants turn light energy into chemical energy, packaging carbon dioxide from the air into sugars. Those sugars are not only fuel; they are building material and, crucially, a form of currency. A plant spends them to grow new tissue, to maintain old tissue, and to run an underground economy: exuding compounds into soil, feeding symbiotic fungi, and repairing the constant wear of living.

A shaded seedling has all the same expenses as a sunlit tree, but far less income. Forest ecologists call this a *source—sink* system. A *source* is a part of the system that produces a surplus of carbohydrates (often the sunlit crown of a mature tree), while a *sink* is a part that consumes more than it produces (often roots, developing cones, new leaves, and, under deep shade, an entire small plant). In everyday terms, the canopy runs a profit; the understory runs a deficit.

That much is uncontroversial. The more intriguing question is whether those deficits can be partly covered by carbon arriving from elsewhere—not via falling needles and decaying logs over years, but as sugars moving within a living network over days and weeks. The notion has become famous under a nickname: the idea that a large tree acts as a *mother* to smaller plants, sending them food underground. The metaphor is compelling because it matches our social instincts. Yet forests do not run on sentiment. If carbon moves between plants, it does so by the physics and physiology of gradients, plumbing, and biological trade-offs. The challenge is to explain *how* movement could occur, and then to ask *how much* it matters.

The first step is to understand the interface where any such transfer must pass. Many forest trees form ectomycorrhizas, a partnership in which a fungus wraps the root tip in a sheath and threads between the outer root cells to make a lattice called the Hartig net. (The fungus does not enter the cells themselves; it builds a contact zone around them.) Through this interface, sugars flow from plant to fungus, and nutrients such as nitrogen and phosphorus flow from fungus to plant. The fungus then extends outward into soil as hyphae, fine threads that can explore pores too small for roots and can fuse with other hyphae, sometimes linking the root tips of different plants.

To picture this as a network is accurate in a limited but important sense: a continuous fungal body can physically connect multiple roots. The more demanding claim is that carbon moves from one plant, into the fungus, and then onward into another plant in a way that can be detected and that influences survival.

How does one prove that? The difficulty is not philosophical; it is practical. Soil is a confusing place. Carbon compounds leak from roots; microbes consume and release carbon; fungi store carbon; and plants can take up carbon compounds directly from soil. If one seedling later contains carbon that can be traced back to a neighbour, was that carbon delivered through fungal hyphae like water through a hose, or did it simply leak into the soil and get reabsorbed? Distinguishing those routes is where the story becomes both ingenious and, at times, contentious.

A turning point came when researchers began using isotopes as tracers. Carbon comes in different isotopic forms. The most common, ^{12}C, has six protons and six neutrons. A small fraction is ^{13}C, heavier by one neutron. Plants discriminate slightly between these isotopes during photosynthesis, but the key for experiments is that scientists can supply carbon dioxide enriched in ^{13}C (or radioactive ^{14}C) to one plant and then look for that label elsewhere. A labelled atom is like a signed banknote: it can move invisibly, but later it can be counted.

In one of the foundational field demonstrations, Simard and Perry (1997) worked with a familiar Pacific Northwest pairing: paper birch and Douglas-fir. These species often grow together, and their seedlings can spend years under a partial canopy. The researchers labelled one plant with an isotope and tracked the appearance of that label in the other, under conditions where mycorrhizal connections could form. They found evidence of bidirectional carbon movement, with net gain for shaded Douglas-fir seedlings under some conditions. The results were electrifying because they were not confined to a growth chamber: they were rooted in the messy realities of a forest.

Yet even that elegance leaves hard questions. When carbon moves, *what form does it take*? The plant's main export to roots is typically sucrose, but fungi can convert incoming sugars into other compounds, including storage molecules, and can release carbon as exudates. A label in a receiver plant could, in principle, have travelled as a metabolite that is not sugar at all, or it could have travelled indirectly after microbial processing. Because of these possibilities, some of the most careful early work tried to create physical designs that separate hyphal transport from simple diffusion through soil.

Francis and Read (1984), working with arbuscular mycorrhizas (a different major mycorrhizal type, where fungal structures form inside root cells), used autoradiography—essentially photographic detection of radioactivity—to show labelled carbon aligned with fungal hyphae linking plants. One reason this mattered historically is that it answered a skeptical, almost courtroom-style objection: *show the route*. It is harder to argue for pure leakage-and-uptake when the signal sits along a visible fungal bridge.

The forest, however, is not a laboratory diagram. Even if carbon can move, its ecological meaning depends on magnitude. A mature tree's carbon budget is vast; a seedling's is tiny. That asymmetry tempts us to assume that even a small subsidy could be decisive. But small in what units? A result might be a large *fraction* of the receiver's photosynthetic uptake on a cloudy day, and still a minuscule *amount* relative to the donor's daily production. The interpretation depends on scale.

One useful anchor is the idea of *carbon source strength*. A plant in strong light has high source strength: it can feed itself and pay its fungal partners. A plant in deep shade has low source strength: it struggles to pay its bills. Many experiments that show increased carbon transfer into a receiver plant do so by manipulating shade. This is not an incidental detail; it points to mechanism. A shaded plant becomes a stronger sink. If a connection exists, gradients intensify. In that sense, the most parsimonious explanation for much of the observed carbon movement is not parental care but plumbing: carbon follows demand, constrained by pathways that fungi help build.

A fungal network is not a passive pipe, though. It is a living organism with its own accounting. The fungus receives carbon, spends it on growth and maintenance, and acquires nutrients that can be traded back to plants. The tempting analogy is a marketplace: plants pay carbon; fungi pay nitrogen and phosphorus; both adjust their behaviour to maximize return. Some experimental evidence supports aspects of this idea, especially for nutrient allocation. Fellbaum and Mensah (2014), for instance, showed that arbuscular mycorrhizal fungi preferentially sent phosphorus toward hosts with stronger carbon supply—the well-lit plants—while still maintaining some colonization and supply to shaded hosts. That pattern looks like differential investment: the fungus rewards the better payer.

But carbon transfer between plants is not automatically explained by a market story. Bunn and Corrêa (2024) emphasize that carbon flow to fungi often reflects basic physiological control and stoichiometry rather than deliberate pricing: plants allocate carbon in ways shaped by their own growth, nutrient status, and constraints, and fungi respond to carbon supply. In other words, both partners are strategic in the broad evolutionary sense, but the day-to-day flow may be more like a gradient-driven circulation than a negotiated contract.

To make this concrete, imagine a sunlit hemlock and a shaded hemlock seedling connected by the same fungal individual. The fungus receives carbon from the sunlit host, and some of that carbon is used to extend hyphae through soil, maintain enzymes, and store energy. Those hyphae also deliver nitrogen and phosphorus. The shaded seedling, being carbon-poor, may offer little carbon to the fungus but may still be colonized because the fungus benefits from occupying multiple hosts, hedging its bets against the death of any one. If the seedling is colonized, it gains improved nutrient and water access, which can matter even when carbon is scarce.

Where does interplant carbon transfer fit? One plausible route is that carbon enters fungal tissues from the donor and a fraction moves through hyphae toward another root tip simply because the fungal body is continuous and resources are translocated internally. If the fungus is delivering nutrients to the seedling's root interface, it may also deliver carbon-containing compounds or leak them locally, which the seedling could absorb. Another route is that the fungus, fueled by donor carbon, supports a larger hyphal system that increases nutrient capture for *both* plants; the seedling then grows more because of nutrients, not because of imported carbon. In that case, isotope transfer might still occur, but the main ecological benefit would be nutritional rather than energetic.

A third route is less romantic but often overlooked: the soil itself as a mixing bowl. Root exudation is real; microbial turnover is rapid; dissolved organic carbon moves; and plants can take up small organic molecules. If donor-labelled carbon leaks into soil and is then absorbed by the receiver, the label still moves between plants, but the fungal network is no longer the essential conduit. This is why experimental designs frequently use mesh barriers that allow hyphae through but restrict roots, or they manipulate fungal connectivity while keeping other variables constant.

One of the more sobering lessons from this literature is that the *existence* of transfer and its *importance* are different claims. The first is now well supported in multiple systems: labelled carbon can be detected moving into neighbouring plants under conditions that allow mycorrhizal connection, and shading often strengthens the signal. The second claim—that such transfers are a major subsidy that routinely determines who survives in the understory—is harder. In many forests, the decisive limiting factors for seedlings include light, herbivory, pathogens, and moisture; carbon is the common currency through which these pressures are paid, but the network is only one part of the balance sheet.

It helps to keep the time scales straight. A mature tree stores carbon in wood for decades. A seedling balances its carbon budget day by day. A small amount of transferred carbon might not build much new wood, but it might pay for maintenance respiration during a bad stretch: keeping cells alive, repairing membranes, maintaining ion gradients. In that sense, a tiny trickle could buy time. Whether that time turns into established growth depends on what happens aboveground: a gap opens, light arrives, and the seedling suddenly has its own source strength.

This is where the understory's drama becomes visible. Foresters have long known that many conifer seedlings can *persist* under shade for years, scarcely growing, waiting for opportunity. The classic picture is often explained purely in terms of low metabolic demand and shade tolerance. Mycorrhizal connections add a nuance: persistence may be supported not only by the seedling's own frugality but by its integration into a larger belowground system that stabilizes nutrient supply and, in some contexts, provides a modest carbon supplement. Even if carbon transfer is rarely the main meal, it may be the difference between fasting and starvation.

The evidence for this kind of subtle support is one reason the debate has been so animated. When a story is subtle, metaphors rush in. The *mother tree* narrative gained traction because it offered a human-scale interpretation of an invisible process. But the more instructive metaphor may be economic rather than familial. The forest is full of unequal exchange simply because light is unequal. A canopy tree has access to an energy source a seedling does not. Carbon follows light; fungi follow carbon; nutrients follow hyphae; and seedlings follow whatever makes persistence possible.

A cross-section sketch can make these relationships legible without turning them into morality tales. The critical features are a sunlit source, a shaded sink, the mycorrhizal interface at root tips, and the idea that most carbon stays within each plant while a smaller, context-dependent fraction can pass through fungal tissues or leak into the shared soil pool. Isotopes, in this picture, are not magic; they are the dye that makes the plumbing visible.

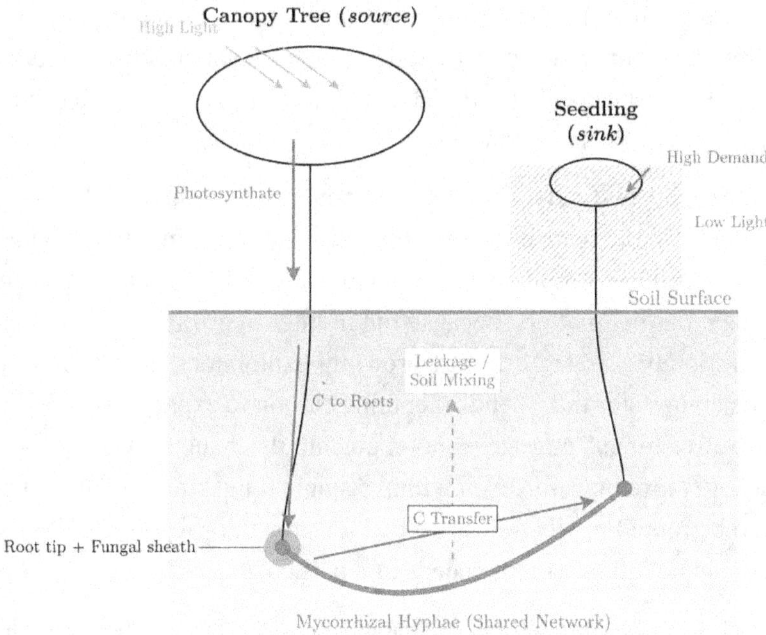

Tracer Idea: Label donor canopy with $^{13}CO_2$ or $^{14}CO_2$, then detect label in hyphae and receiver tissues.

A diagram like this risks making the system look too neat. Real soil is lumpy. Hyphae do not form a single cable but a braided, branching mesh; roots are colonized by multiple fungal species; and connections can be transient. Still, the picture captures the essential logic: carbon is produced where light is abundant, and any movement into the shade must be driven by differential supply and demand. The fungal sheath at the root tip is the customs checkpoint. Nothing moves between partners without passing through that intimate interface.

One reason the isotopic experiments matter is that they shift the conversation from metaphor to measurement. A label can be counted. It can be found in the fungal tissue itself, strengthening the case that the fungus is not merely an onlooker. It can be tracked over time, showing whether transfer is a brief pulse or a sustained trickle. And it can be compared across treatments—shade versus sun, connected versus disconnected, fertilized versus unfertilized—to reveal what actually controls the flow.

Those controls tend to look less like benevolence and more like physiology. Shade increases sink strength. Nutrient limitation changes how much carbon plants allocate belowground. The identity of the fungal partner matters, because fungi differ in growth form, storage, and foraging strategy. A fast-growing, exploratory fungus may act like an investor that spends incoming carbon to expand; a more conservative fungus may store more, potentially changing what is available to move onward. Even within a single fungal individual, carbon can be routed to where it is needed for growth or repair, which may or may not align with the needs of a particular seedling.

For the shaded plant, then, the underground connection is best understood as a conditional opportunity. When light is scarce, carbon is precious and growth is constrained. Any mechanism that stabilizes the budget—more reliable nutrients, access to microsites of moisture, even a small influx of labelled carbon—can translate into persistence. Persistence is not a sentimental outcome; it is a demographic one. A seedling that survives for ten years under shade is present when a branch falls and a window of light opens. Many forest canopies are replaced not by a burst of germination but by a quiet bank of waiting juveniles.

The most honest conclusion is also the most interesting: forests can be both interconnected and unsentimental. Carbon transfer through mycorrhizal links is real enough to be detected, and it is shaped by the stark inequality of light. Yet the network does not abolish competition or rewrite the rules of photosynthesis. It offers pathways and constraints—channels through which materials *can* flow when gradients and circumstances make it worthwhile. In the shade, life is negotiated in small sums: a few molecules here, a few hours of respiration there, a marginal improvement in nutrient status that keeps a bud alive through winter. The understory's patience is built from such margins, and the fungal web is one of the quiet mechanisms that makes those margins possible.

5.2 Hydraulic Redistribution: The Water Brokers

A drought does not arrive as a single event in a forest; it arrives as a gradient. The surface soil dries first, sometimes within a few clear days. A few centimetres down, moisture lingers longer, protected from wind and sun. Deeper still, especially where roots can reach into fractured rock or a buried lens of finer soil, water may remain available long after the understory has begun to crack and crisp. Plants experience this not as a calendar date but as a set of tensions: water held tightly in small pores, water pulled upward through stems, water evaporating from leaves. When ecologists say that water moves from wet to dry, they mean from high to low *water potential*—a measure of how strongly water is held, and how strongly it is being pulled.

Water potential is the quiet dictator of plant life. It explains why water can move upward, against gravity, inside a tree: evaporation from leaves creates a pull that draws a continuous column of water through xylem conduits. It also explains why, at night, when stomata close and transpiration slows, a deep-rooted tree can become a kind of subterranean pump in reverse. If deep roots sit in moist soil and shallow roots sit in dry soil, water can flow *out* of the root system into the parched upper layers. This is hydraulic redistribution. When the direction is upward, the most famous form is *hydraulic lift*.

The phrase has a charming Victorian solidity, as if a forest has hidden machinery. But the mechanism is brutally simple. Roots, like any wet membrane connected to a wet network, allow water to move along a gradient when the plant is not actively pulling water toward the canopy. Deep soil water, at relatively high water potential, can enter roots. If the upper soil is drier (more negative water potential), water can then leak back out of shallow roots into that dry zone. The plant does not need to decide; the physics does it.

Foresters noticed the *effects* long before anyone had the vocabulary. In dry summers, the surface soil beneath large trees can remain surprisingly moist compared with open ground nearby, even when no rain has fallen. Seedlings often survive better in the shadow and vicinity of established trees than in exposed microsites, sometimes despite competition for light. Part of that survival is shade and temperature. Part is soil structure and organic matter. Part, increasingly, is understood as the nocturnal rewetting of the upper soil—a subsidy of water moved from depth.

That story would already be fascinating if it involved only roots. The twist is that roots are not alone in the soil. They are entwined with fungi, and those fungi are not merely passengers. A mycorrhizal fungus is a living extension of a root system, with hyphae fine enough to enter pores that roots cannot, and long enough to bridge between microsites that differ in moisture by millimetres. Once one accepts that hyphae are continuous filaments filled with cytoplasm and bounded by membranes, it becomes hard to avoid the question: can fungal networks move water too?

The answer, supported by a growing set of experiments, is yes—sometimes. But the most useful way to say it is that fungal networks can act as *water brokers*: not the original source of water, and not the only route, but an additional pathway that can connect wet and dry domains, sometimes linking plants that would otherwise be isolated from one another's hydraulic influence.

The cautious tone matters. The idea of an underground network that shares resources is irresistibly anthropomorphic. Water-sharing, in particular, invites a sentimental picture: a mature tree "giving" water to a seedling. Yet water is not like carbon. A tree spends carbon to build and maintain tissues; water mostly passes through. Moving water through hyphae does not necessarily cost much carbon directly, but it can carry risks: it can hydrate competing roots, stimulate microbes that consume nutrients, or create conditions favourable to pathogens. For fungi, moving water could enable hyphal growth and enzyme function, but it could also make their habitat more hospitable to competitors. Any general statement about benefits and altruism dissolves quickly into context.

It helps to start from the smallest scale: a hypha as plumbing. A fungal hypha is typically only a few micrometres thick. Inside it, water can move by bulk flow when there is a pressure gradient, and it can move by diffusion when gradients are small. Many fungi also form thicker cords or rhizomorphs, bundled structures that function like highways for transport. If a cord spans a dry patch of soil, it can, in principle, deliver water to hyphae at its far end, keeping them active where they would otherwise desiccate.

This is not speculative; it is something one can see in the laboratory. Place a fungal cord across two soil compartments, one wet and one dry, and the fungus can maintain growth into the dry side far longer than a purely local water supply would allow. That observation alone, though, does not yet say anything about plants. The challenge is to show that water moved through hyphae can be delivered to a plant root or can alter plant water status.

To do that, researchers have borrowed the same logic used for carbon: tracers and barriers. Water can be labelled with deuterium (^{2}H), a stable isotope of hydrogen. Add deuterium-enriched water to a donor zone, keep the receiver zone free of that label, and then look for deuterium in the receiver plant's transpiration stream. If the label appears when roots are excluded but fungal hyphae can pass, the case for a hyphal pathway becomes compelling.

Egerton-Warburton and Querejeta (2007) provided an influential demonstration along these lines. Using experimental designs that separated roots while permitting fungal connections, they detected deuterium in the water transpired by receiver plants and used dyes to visualize movement through fungal structures. Their work was important not only because it showed that transfer could occur, but because it showed that the transport path could be constrained to hyphae rather than roots. The results also carried a warning that has echoed through subsequent studies: the fraction of the fungal network actively involved in water transfer may be small. A mycorrhizal system contains an enormous length of hyphae, but only certain strands may act as effective conduits at a given moment, depending on soil contact, continuity, and the dynamic hydration state of the fungus.

Field studies have added an extra layer of realism. In a ponderosa pine system, Warren and Brooks (2008) used manipulations that limited root access while allowing fine fungal connections, and they traced water movement consistent with an ectomycorrhizal pathway contributing to hydraulic redistribution from mature trees toward seedlings. Forest fieldwork rarely yields the clean lines of a laboratory graph. Soil is heterogeneous; roots and hyphae are everywhere; and any barrier changes the microenvironment. Yet the value of such studies lies in showing that hyphal water movement is not merely an artefact of pots: it can plausibly contribute under real drought conditions where shallow soil becomes a hydraulic bottleneck for seedlings.

If that is the plant-fungal version, there is an even more surprising extension: fungi can redistribute water even without plants. Guhr and Borken (2015) demonstrated that a saprotrophic fungus—a decomposer that does not form mycorrhizas—could move water across a steep gradient between soil compartments. The ecological consequence in their system was dramatic: by bringing moisture into an otherwise dry zone, the fungus triggered a surge in enzyme activity and carbon mineralization. This result matters because it untethers the concept from plant mutualism. Water redistribution is not a special favour done by mycorrhizas; it can be a general property of fungal networks as living filaments occupying patchy soil.

For forests, that generality is both exciting and unsettling. It means fungi are not only passengers on plant hydraulics but potential engineers of the soil's own water map. A small increase in moisture around a root tip can make the difference between a nutrient ion being mobile or stuck; between microbial dormancy and respiration; between a pathogen's spore remaining inert or germinating. The fungus that moves water may therefore be affecting plant survival indirectly through changes in the microbial community and nutrient cycling, not only through the direct delivery of water to a seedling.

A concrete example makes the stakes clearer. Consider a late-summer drought in a mixed conifer forest. Mature trees with deep roots still access moisture, but surface soils where seedlings live have dried to the point that water is held at very low potential. A seedling's fine roots, close to the surface, are like straws pushed into a nearly empty sponge. Every day, the seedling loses water through its needles; every night, it tries to recover. If a mycorrhizal network connects the seedling's root tips to hyphae extending into slightly wetter microsites, those hyphae can act as capillaries that keep the immediate root zone from reaching lethal dryness. This is not necessarily a large volume of water. A seedling does not need to be flooded; it needs to keep living tissues above a threshold where membranes remain intact and the plant can reopen stomata when conditions improve.

Here the word *buffer* is more honest than *subsidy*. The fungal pathway may not supply enough water to sustain transpiration like a deep root would. Its contribution may be to smooth extremes: slowing the rate at which the rhizosphere—the thin zone of soil influenced by roots—dries out, and speeding the rate at which it rehydrates at night. That smoothing can matter disproportionately because plants often fail at thresholds. A slight delay in reaching the water potential that causes xylem cavitation (the formation of air bubbles that block water flow) can be the difference between recovery and death.

This kind of threshold thinking also explains why results across studies can be variable. Hydraulic redistribution by roots is now documented in many ecosystems, but its magnitude varies with soil texture, rooting depth, and atmospheric demand. Add fungi, and variability increases. The fungal network may be continuous or fragmented. Hyphae may be active or dormant. Some fungi produce more rhizomorphs, potentially supporting longer-distance transport, while others are more diffuse. Hyphae may be insulated by soil aggregates or exposed to air-filled pores that sever water films. Even the direction of benefit can shift. A fungus that moves water into dry soil could inadvertently help neighbouring plants that compete with its host, diluting the host's advantage. A host might tolerate that if the fungus also improves nutrient uptake, or if the host's deep water access is ample. But in a harsher drought, that same dilution could be costly.

A historical anecdote captures the scientific mood around these ideas. In the early decades of mycorrhizal research, the dominant story was nutritional: fungi help plants obtain phosphorus and nitrogen in exchange for carbon. Water was considered a side benefit at most, an indirect consequence of improved root health and soil structure. The turn toward hydraulics emerged as ecologists studying drylands and seasonally dry forests began to treat water not as background but as the main currency of survival. Hydraulic lift, once controversial, became accepted as both widespread and ecologically meaningful. It was almost inevitable that someone would ask whether the abundant fungal filaments in those same soils were participating. The fascination was not just that fungi might move water, but that they might connect water movement to the social life of plants: one plant's deep access could influence another plant's surface conditions through a shared fungal mesh.

Keeping that fascination disciplined requires a clear mental model of pathways. There are at least two plausible ways a seedling might benefit from a deep-rooted neighbour during drought. One is purely root-mediated: the mature tree lifts water into the shallow soil, and the seedling's roots drink it directly. The other is hypha-mediated: the mature tree's lifted water hydrates the fungus associated with it, and that fungal network transports water, or maintains moisture films, toward the seedling's rhizosphere. In reality, both could operate simultaneously, and separating them experimentally is difficult.

Mesh barriers help, but they are never perfect imitations of natural soil architecture.

This is also where language can mislead. Calling the fungus a *pipeline* suggests a rigid tube. Many hyphae are more like wet threads pressed through a granular medium. Water can move along the hypha's interior, but it can also move along its surface as a thin film, especially in soils where water exists mainly as films around particles. In a drying soil, the continuity of those films breaks, and movement becomes discontinuous. A hypha can sometimes maintain continuity across air gaps in a way that soil films cannot, effectively bridging a hydraulic break. That bridging might be as important as any bulk transport inside the hypha.

The saprotrophic example makes another point: moving water is not necessarily about helping plants; it can be about keeping fungal metabolism running. Many fungal enzymes involved in decomposition require a minimum level of hydration to function. If a fungus can bring water into a dry patch of soil, it can continue decomposing organic matter there, liberating nutrients. Those nutrients, in turn, may become available to plants, including mycorrhizal hosts. Under drought, then, water redistribution could influence forest nutrition indirectly by keeping decomposition and nutrient cycling alive in microsites that would otherwise shut down.

There is a human resonance here, because forests under drought are no longer an occasional curiosity. In many regions, warmer temperatures increase evaporative demand, and droughts become more intense even when rainfall totals do not change dramatically. Seedlings are often the first casualties. A mature tree can survive a dry year by drawing on deep reserves and stored carbon; a seedling has no such buffer. If fungal networks can slightly increase the probability that seedlings endure through a critical window, they can influence regeneration patterns, and regeneration patterns determine the future forest.

This does not mean fungal hydraulics will save forests. Drought kills by multiple routes: hydraulic failure, carbon starvation, heat stress, pest outbreaks. A hyphal bridge cannot compensate for a complete loss of soil moisture or for a heatwave that forces stomata closed for weeks. Yet the notion of fungi as water brokers changes how one imagines resilience. Resilience is not a single trait possessed by a species; it is an emergent property of many small processes that reduce the chance of catastrophic failure. Some of those processes are aboveground, like shading that lowers leaf temperature. Some are within plants, like the ability to resist cavitation. Some are between organisms, like the fungal filaments that keep a root tip from reaching the point of no return.

The most compelling picture is not of water being generously handed from one plant to another, but of a forest floor held together by countless microscopic connections that make drought less binary. Wet and dry are not states; they are gradients. Life persists by finding paths across gradients. A fungal cord running from a moist pocket of soil into a dry one is a path; a mycorrhizal sheath around a root tip is a path; the thin film of water clinging to a hypha's surface is a path. In the evenings of a long summer drought, when the air cools and transpiration quiets, those paths can briefly widen, letting water slip into places that would otherwise stay closed.

By morning, the sun returns, and the gradients steepen again. The lifted water begins to be spent, evaporated leaf by leaf into the atmosphere. What remains is not a gift but a redistributed arrangement of moisture in the soil, a small rearrangement with outsized consequences for the organisms living at the edge of desiccation.

5.3 Chemical Telephony and Defense

When a caterpillar begins feeding on a leaf, the first injury is small: a ragged edge, a smear of saliva, a few ruptured cells. The response is disproportionately large. Within minutes, electrical changes ripple across the wounded tissue. Within hours, the plant's chemistry shifts: bitter compounds rise, proteins that gum up insect digestion appear, and volatile scents drift into the air, sometimes summoning predators and parasites that treat herbivores as dinner. A tree or herb cannot run from an attacker, so it makes itself difficult to eat and recruits others to do the eating.

This defensive sophistication is no longer surprising. It is one of the quiet revolutions in modern plant biology, and it has a history. In the early 1980s, two entomologists, Ian Baldwin and Jack Schultz, helped make plant defence feel less like an anecdote and more like a system, showing that damage could induce chemical changes and that those changes could affect herbivores. Around the same time, researchers began to take seriously the idea that plants use airborne compounds as signals, not just as metabolic exhaust: a wounded plant releases volatile organic compounds (VOCs), and nearby plants can respond by priming their own defences. The popular version of that story is tidy: plants smell trouble and prepare.

The underground version is less tidy and, for that reason, more interesting. If fungi connect roots, and if sugars and water can sometimes move along those connections, could warning signals do the same? Could a plant under attack alert an untouched neighbour through a shared mycorrhizal network, so that the neighbour becomes less palatable before the herbivore arrives? This is the claim behind the headline phrase "talking trees," and it is where metaphor begins to outrun mechanism.

To approach it without either cynicism or credulity, it helps to define what counts as a signal. In biology, a signal is information that changes behaviour in a way that is, on balance, beneficial for the sender, the receiver, or both. That definition is already a trap, because it invites intentional language. Plants do not intend; fungi do not intend. But both have evolved traits that change others' physiology. A compound released from a wounded root might be a cue (something other organisms can exploit), or it might be a signal (something shaped by selection to influence others). Experimentally, the first step is not to decide which it is; it is to show that something moves, and that it produces a reproducible response.

The best-studied candidates are plant hormones and hormone-like molecules involved in defence. Jasmonates (including jasmonic acid and its derivatives) are central to defence against chewing insects. Salicylic acid is central to defence against many pathogens and some piercing-sucking insects. Plants also use peptides, reactive oxygen species, calcium waves, and electrical changes. Any of these could, in principle, travel belowground.

Yet the soil is a noisy medium. Roots leak chemicals; microbes transform them; fungi absorb and secrete compounds; and the physical space is a mosaic of air pockets, water films, and sticky surfaces. If a receiver plant shows increased defence, it might be because a compound travelled through fungal hyphae. It might also be because microbes moved, because the fungus changed its colonization behaviour, because the plants competed for nutrients in a way that altered defence investment, or because VOCs drifted through the air while researchers were focused on the soil. The burden of proof is therefore higher than it first appears.

One of the most vivid demonstrations that spurred interest came from work on broad bean plants and aphids. In a widely discussed study, Babikova and colleagues (2013) connected plants via common mycorrhizal networks and exposed one plant to aphid attack. Neighbouring plants that were not attacked showed changes in their volatile emissions, including increased methyl salicylate, a compound that can repel aphids and attract aphid enemies. The striking point was that the effect was observed when plants were connected belowground, suggesting a role for the mycorrhizal pathway. In a system where aphids reproduce quickly and can move between plants, early warning would be valuable: it is easier to deter a pioneer aphid than an established colony.

A year later, Song and colleagues (2014) offered another influential example using tomato plants and caterpillars (Spodoptera litura). Their design linked donor and receiver plants via mycorrhizal fungi and then damaged the donor by herbivory. Receiver plants showed increased expression of defence-related genes and enzymes and became more resistant to insect feeding. The cleverest part was genetic: when the donor plant lacked a key capacity to produce jasmonate signals (a mutant defective in jasmonic acid biosynthesis), the receiver's induction was abolished. That result did not prove that jasmonate itself travelled through the hyphae, but it tied the phenomenon to jasmonate-dependent signalling in the donor. It narrowed the story from "something about damage affects neighbours" to "a jasmonate-dependent pathway is necessary."

These studies make it difficult to dismiss the idea outright. Plants connected by fungi can show induced or primed defences in the absence of direct attack. The responses can include changes that matter ecologically: altered volatile bouquets, altered herbivore performance. At the level of phenomenon, something real is happening in these experimental systems.

The harder question is *what* is moving and *how*. One seductive assumption is that a defence hormone simply flows from an attacked plant, through the fungus, into its neighbour, like dye through a straw. But hormones are active at very low concentrations, and they are often rapidly modified or broken down. A molecule that acts as a signal inside a plant is not automatically stable in soil or compatible with fungal transport. It also may not need to be transported in its final form. A plant could export a precursor, a fragment, or an entirely different compound that triggers hormone production in the neighbour.

There is also another possibility: the fungus itself may be responding to the attacked plant and then altering its behaviour toward other plants. Fungi are not inert wires. They can change gene expression, nutrient allocation, and growth direction in response to their environment. If a host becomes stressed or begins to allocate carbon differently, the fungus may shift its investment among its connected hosts. That shift could change nutrient status in the receiver plant, and nutrient status can influence defence. A receiver that gains a little nitrogen or phosphorus might invest more in defensive proteins; alternatively, a receiver that becomes nutrient-limited might reduce growth and allocate proportionally more to defence. In either case, the receiver's defence changes, but the moving entity is not necessarily a hormone signal; it could be a change in resource distribution.

This is why review papers on the topic often read like careful courtroom summaries. Johnson and Gilbert (2015), for example, framed mycelial networks as potential conduits for defence cues while emphasizing how much remains uncertain: the identity of the molecules, the distances involved, the generality across ecosystems, and the fitness interests of fungi. The review's tone is not dampening; it is clarifying. There is a difference between "plants can respond to neighbours through fungal connections" and "trees communicate intentions."

The distinction matters because communication is expensive in evolutionary terms. A signal that reliably benefits receivers can be exploited by cheaters. If a plant could obtain defence priming from a neighbour without paying a cost, it might evolve to lean on others. But plants and fungi are tightly costed organisms. Defensive chemistry often reduces growth. Mycorrhizal carbon payments are substantial. A fungus that indiscriminately spreads warning signals might help a host's competitor. Natural selection rarely favours indiscriminate generosity unless there is a compensating benefit.

It is therefore more plausible to think of these networks as *shared infrastructure* in which multiple processes can be coupled. When one plant is attacked, its metabolism shifts. That shift changes carbon flow to roots. Carbon flow changes fungal physiology. Fungal physiology changes nutrient allocation, water relations, and perhaps microbial communities. Those changes can prime or induce defences in neighbours. In this picture, there can be information transfer without a single, simple "message molecule." The signal is a pattern of correlated changes transmitted through a shared system.

A concrete, almost domestic analogy can help without anthropomorphism. Imagine an apartment building with shared plumbing and heating. If one tenant runs hot water continuously, pressure drops elsewhere; radiators cool; pipes knock; neighbours adjust. No one has "communicated," but information about one apartment's behaviour has propagated through the system, and others respond. A mycorrhizal network could function similarly: not as a telephone line with discrete words, but as a coupled medium in which stress in one host alters conditions experienced by another. The receiver's response can still be adaptive, because the correlation between the neighbour's stress and the receiver's future risk can be real.

The strongest experimental designs try to simplify that coupled system to isolate a transport route. That is why researchers often use paired plants connected by a single fungal species in controlled conditions, sometimes with physical barriers that prevent root contact but allow hyphal passage. They also try to exclude airborne VOC signalling by enclosing shoots separately or filtering air. Even then, alternative explanations persist, because microbes and dissolved compounds can pass through pores and along water films.

This is where the fascination with *electrical* signalling has entered the story. Plants generate electrical signals: wound potentials and action potential-like events that travel through tissues, often coupled to calcium waves and hormonal cascades. Fungi also display electrical phenomena, including membrane potential changes and propagating pulses under certain conditions. In simplified laboratory settings, electrophysiological signals have been recorded moving along mycelial bridges after wounding, suggesting that fungi could, in principle, conduct a rapid signal that is not dependent on slow diffusion of chemicals. If such conduction occurs in natural soils, it would offer a different kind of mechanism: a fast alert that could prime a neighbour before chemical concentrations build up.

The keyword is *if*. Laboratory evidence that a mycelial bridge can conduct electrical changes does not automatically translate to forest conditions. Soil is electrically complex; hyphae are embedded in heterogeneous media; and the architecture of networks is dynamic. The significance may lie less in a literal fungal "nervous system" and more in the idea that signalling may be multimodal: chemical cues, hydraulic changes, nutrient shifts, and electrical dynamics could interact, with fungi acting as integrators rather than mere conduits.

This is also where the phrase "talking trees" becomes misleading. Talking implies intentionality, a sender crafting a message for a receiver. The evidence supports something subtler: plants and fungi can be coupled in ways that allow stress in one to alter physiology in another. That coupling can lead to measurable defensive changes. Whether the evolutionary function is "communication" depends on whose fitness is improved and under what conditions.

The fungal perspective is often neglected in popular accounts. A fungus benefits when its hosts survive and provide carbon, but it also benefits when it can maintain access to multiple hosts. It may be in the fungus's interest to keep several plants alive at moderate performance rather than to maximize the success of a single host. If a warning-induced defence response reduces herbivory on connected plants, the fungus gains indirectly by stabilizing its carbon supply. Alternatively, if inducing defence costs plants growth and therefore carbon allocation, the fungus might not benefit. The fungus could even modulate signal transfer, enhancing or dampening it depending on its own state. That possibility is difficult to test, but it follows naturally once fungi are treated as agents in the ecological sense: organisms with their own constraints and adaptive interests.

The limits of signalling become clearest when one asks about distance and specificity. In a forest, a mycorrhizal network can be extensive, but it is not a single, universal grid. Different plants host different fungi; networks overlap; and connections may form and dissolve. A signal that spreads readily through a laboratory network might attenuate quickly in the field. Even if it spreads, the receiver might be connected to multiple donors, each with different stresses. The receiver then lives in a chorus of cues. Some will be relevant, some false alarms. Overreacting is costly. Underreacting is costly. Natural selection should favour response rules that work on average, which may produce inconsistent outcomes in any given experiment.

There is also the question of timing. Herbivore attacks can escalate quickly. Aphids can colonize in days. Caterpillars can skeletonize leaves in hours. A warning system is only useful if it is faster than the spread of damage. Airborne VOC signals can be fast and can travel metres to tens of metres, but they are subject to wind and dilution. Underground signals, if they exist, may be more protected from wind but slower if they rely on diffusion. A hyphal pathway could offer directed transport, yet that transport is constrained by the fungus's internal flows and the physical state of the soil. During drought, when soil water films break, chemical diffusion slows, but electrical or hydraulic coupling might persist along hyphae that maintain internal continuity. Paradoxically, the conditions that make aboveground signalling unreliable might make belowground pathways comparatively more important, even if the overall magnitude is small.

A final caution comes from a familiar problem in biology: correlation masquerading as communication. Plants growing near each other often share the same soil conditions, microbes, and microclimate. If one plant is attacked, it might be because that spot has higher herbivore pressure; neighbouring plants are therefore at higher risk regardless of any signalling. They may also share similar microbial communities that prime defence. In such cases, a receiver's heightened defence is not caused by a message from a neighbour, but by shared exposure to the same ecological context. The best experiments control for this, but field systems always contain hidden couplings.

The most humane way to end up neither enchanted nor dismissive is to treat mycorrhizal networks as *channels of possibility*. They can, under certain conditions, carry molecules, water, and perhaps electrical changes. They can link organisms into a shared physiological arena where one plant's stress alters another's options. This can look like warning and preparation, and sometimes it may function as such. But it is not a woodland parliament exchanging speeches. It is an entangled set of living tissues responding to gradients, constraints, and the immediate arithmetic of survival.

In a forest, defence is not only about fighting; it is about timing. A plant that is slightly primed may respond a little faster, accumulate defensive compounds a little earlier, and suffer a little less damage. Those "littles" matter because herbivores and pathogens exploit delays. A mycorrhizal connection that turns a delayed response into a timely one does not need to transmit a poetic message. It needs only to make the receiver's chemistry begin before teeth arrive, and to do so often enough that the strategy pays its costs.

By late summer, when leaves are toughened and insects are nearing the end of their feeding season, the forest's defensive chemistry has already written itself into the air: faint green odours, resinous notes, the sharpness of crushed needles. Beneath the litter, where roots and hyphae press through dark pores, the signals are quieter and harder to isolate. Yet quiet is not the same as absent. In that darkness, a wounded plant can change the shared conditions of the rhizosphere, and a neighbour can respond as if it has overheard something worth acting on.

Chapter 6

Detecting the Ghostly Signal

The underground economy operates in silence and darkness, leaving no obvious tracks for human observers. To reveal these covert exchanges, researchers have adapted tools from nuclear physics and forensics, turning atomic weight and genetic scraps into irrefutable evidence of connection.

6.1 Following the Atomic Breadcrumbs

Forests look solid and self-contained aboveground: trunks fixed in place, crowns stitched to the sky. Belowground, the story is messier. Roots leak sugars and acids; microbes eat, transform, and repackage them; fungi thread through pores in the soil that a root hair cannot enter. If one tree's carbon ends up inside another tree's tissue, how could anyone *prove* the route it took?

The difficulty is not simply darkness. It is that carbon and nutrients are restless. A tree exhales carbon dioxide through respiration; leaves and bark shed fragments; roots slough cells; rainwater shuttles dissolved compounds; microbes metabolize and re-release them. In that churn, a scientist needs a marker that behaves like the substance of interest but carries a signature so distinctive that it cannot be waved away as coincidence. Stable isotopes provide exactly that sort of signature: an atomic breadcrumb trail.

An isotope is a version of an element with the same number of protons but a different number of neutrons. Carbon, for example, almost always comes as carbon-12: six protons and six neutrons. A small fraction comes as carbon-13, with an extra neutron, making it slightly heavier. Both are stable, meaning they do not decay radioactively. Chemically they act nearly identically: a carbon atom still forms four bonds; it still fits into sugars, starch, cellulose, and the countless molecules that make a tree a tree. But physics can tell them apart. The extra neutron changes the atom's mass, and that tiny difference becomes measurable with instruments sensitive enough to weigh molecules by the subtlest increments.

If this sounds like a trick too delicate for a forest, that is because it is. It depends on the fact that life is built from repeated units. A leaf's carbohydrates are made by photosynthesis from carbon dioxide. If the carbon dioxide is enriched in carbon-13, the resulting sugars become enriched too. Those sugars can move down the phloem, into roots, and into the fungus that is partnered with those roots. If that fungus is connected to a neighboring tree by a web of hyphae, and if carbon moves across that bridge, then the neighbor should show a measurable enrichment in carbon-13 compared with what it would have had otherwise. The key word is *enrichment.* Scientists are not looking for a single labeled molecule; they are looking for a statistically unmistakable shift in the isotopic composition of tissue.

The elegance of the method is that it turns a philosophical problem—how to infer a hidden connection—into a practical one: how to label one organism cleanly, and how to measure the label precisely enough to reject alternative explanations.

One of the most vivid ways to label a tree is disarmingly simple:

place a clear bag over a branch or a small tree and let it breathe a different atmosphere for a short time. In the field experiments that have become classics in this area, researchers enclosed a "donor" seedling or sapling in a plastic chamber and pulsed it with carbon dioxide enriched in carbon-13. For an hour or two, the tree photosynthesizes with this modified gas. Nothing else about the setup changes. The light is the same. The temperature is nearly the same. The tree is still making sugars for its own life. It is only making them from carbon atoms that carry an extra neutron.

After the pulse, the bag is removed and the donor is returned to normal air. The labeled sugars begin to move and to be used. Some are burned for energy, releasing carbon dioxide enriched in carbon-13. Some are stored. Some are sent belowground. If the donor is linked into a mycorrhizal network—a fungal partnership where tree roots and fungal hyphae exchange resources—then part of that labeled carbon may enter fungal tissue. If fungal hyphae connect the donor's roots to another tree, carbon may pass along hyphae or through structures where hyphae meet roots.

Days later, researchers harvest small samples: needles, fine roots, sometimes bits of fungal tissue. They measure the ratio of carbon-13 to carbon-12. That ratio is naturally fairly constant in the atmosphere and in plants, with small, predictable deviations. When a plant has been pulsed with carbon-13, its ratio becomes conspicuously higher. If a neighboring tree shows an increase as well, the simplest interpretation is that carbon from the donor has found its way into the receiver. But science is an art of refusing simplicity too soon. Could the receiver have absorbed carbon dioxide respired by the donor? Could labeled compounds have leaked into the soil and been taken up by roots directly? Could the trees have been connected by root grafts rather than fungi? Each possibility has its own experimental fingerprints, and good isotope studies try to separate them.

The physical idea beneath the measurement is straightforward even if the instrumentation is not. An isotope ratio mass spectrometer takes a sample, converts it to a gas (often carbon dioxide for carbon measurements), and sorts ions by mass. Carbon dioxide made with carbon-12 has a slightly different mass than carbon dioxide made with carbon-13. By counting them, the instrument provides a ratio. Often the result is expressed as a delta value, $\delta^{13}C$, which reports how much the sample differs from a standard. The point for a general reader is not the unit but the confidence: a tiny neutron-level difference can be amplified into a number that can be compared across tissues, trees, and days.

The method gained its cultural power not because it was new—stable isotopes had long been used in geology, physiology, and food-web ecology—but because it offered a forensic approach to a claim that had previously rested on inference. Before isotopic labeling, one could watch seedlings grow better near certain trees and speculate about belowground help. One could see fungal strands on roots and imagine them as conduits. But imagination is cheap. Carbon-13 allowed a kind of honest accounting: if carbon moved, the atoms themselves would testify.

A concrete example helps. In the 1990s, Suzanne Simard and colleagues carried out field experiments in which they labeled trees and then detected the movement of carbon between different species linked by ectomycorrhizal fungi—a form of mycorrhiza common in many temperate and boreal forests where fungi sheath the roots and extend hyphae outward. Their work, including a notable 1997 study with David Perry, used reciprocal labeling to test whether carbon could move in both directions between neighboring trees. The important nuance was context: which tree was in the role of *source* (net exporter of carbon) and which was a *sink* (net importer) depended on conditions like shading. A tree under heavy shade may become carbon-starved, while its neighbor in sun becomes carbon-rich. In that source—sink landscape, isotopes can reveal not only that a connection exists, but that the direction and magnitude of transfer can change with circumstance.

This is where stable isotopes behave like breadcrumbs rather than like a glowing dye. A dye is obvious, but it is also artificial; it can clog pathways, alter behavior, or announce itself to biology as an intruder. Carbon-13 is not a dye. It is carbon. The tree does not know it is being tracked. That is why its testimony is persuasive.

Yet a breadcrumb trail can be smeared by wind and rain. The biggest practical problem in carbon-13 labeling of trees is that plants and soil organisms exchange carbon in more than one way. Leaves take in carbon dioxide from the air and also release it; soil releases carbon dioxide from microbial respiration; roots release carbon compounds into the soil; fungi release enzymes and metabolites. If a receiver tree shows an increase in carbon-13, an investigator must ask: *where did that carbon-13 enter?* A careful experiment tries to ensure that only the donor experiences the labeled carbon dioxide pulse. It also tries to prevent labeled respiratory carbon dioxide from being inhaled by the receiver's leaves. Sometimes this is done by labeling at times of low atmospheric mixing or by physically separating canopies. More often, researchers rely on a combination of timing and magnitude: if enrichment appears first in fungal tissues or in receiver roots rather than in receiver leaves, and if the timing fits known pathways of transport, the case strengthens. Still, the best studies do not rest on timing alone. They include controls designed to break the fungal connection while leaving other routes open, so that any remaining enrichment can be attributed to those other routes.

A stable isotope pulse is not limited to carbon. The same logic applies to nitrogen and phosphorus, elements that often limit forest growth. Nitrogen has two stable isotopes, ^{14}N and ^{15}N, and phosphorus can be tracked with radioisotopes rather than stable isotopes because it has only one stable isotope. In practice, carbon-13 labeling has become emblematic because it integrates so naturally with photosynthesis: a tree is already a machine for turning carbon dioxide into mobile sugars. Label the gas and the tree labels itself.

The "days later" part of these experiments matters. Carbon does not teleport. Sugars must travel down phloem, move into root tissues, and cross membranes into fungal partners. Fungi must then move carbon through hyphae, perhaps incorporating some into their own biomass, perhaps releasing some as metabolites, perhaps delivering some to other roots. Measurements at multiple time points can reveal a wave of enrichment: first in donor leaves, then donor roots, then fungal tissues, then receiver roots and shoots. That sequence becomes a signature of pathway. When researchers use radioisotopes such as carbon-14 in controlled settings, they can even visualize distribution with autoradiography, producing images where darker regions correspond to higher radioactivity. The pictures have a haunting immediacy: a map of where recent photosynthate traveled, not as a metaphor but as a physical trace. Stable isotopes do not produce pictures so easily, but they can be sampled and quantified with fewer regulatory constraints.

Here the method's power meets its philosophical limit. Detecting labeled carbon in a receiver tree is strong evidence of movement. It is not, by itself, evidence of benefit. A teaspoon of sugar delivered to a starving seedling might be lifesaving, or it might be metabolized by microbes before the seedling can use it. A small isotopic enrichment can correspond to a small amount of carbon transfer, or to a larger transfer diluted into a large carbon pool. Interpreting ecological significance requires careful accounting of pool sizes and fluxes, and it requires resisting the temptation to turn "carbon moved" into "carbon mattered." The point of atomic breadcrumbs is not to settle every ecological question. It is to establish the existence and possible pathways of exchange with a rigor that anecdotes cannot provide.

One reason the method is so persuasive to non-specialists is that it resembles detective work. The forest becomes a crime scene in which carbon atoms are witnesses. But unlike television forensics, the forest will not cooperate. Every control has a failure mode. A plastic bag can leak. Labeled air can mix. A barrier intended to block fungal hyphae can leave a seam. Roots can slip under a divider. And because the signals are small, a tiny methodological leak can masquerade as a biological transfer. Good isotope studies therefore have a certain humility built into their design. They rely on replication, on background measurements, on careful calibration of what "enrichment" looks like in the absence of transfer, and on explicit alternative hypotheses. In recent debates about how to interpret common mycorrhizal networks, methodological critiques have emphasized exactly this: carbon transfer is not a fairy tale, but its detection demands an evidentiary standard as strict as any forensic science.

When stable isotopes are used well, they can do something more subtle than provide a yes-or-no answer. They can reveal *directionality*. A reciprocal labeling design can label tree A one day and tree B another day and track enrichment patterns in both directions. They can reveal *contingency*: shading, drought, or damage can shift which tree is a source and which is a sink. And they can reveal *partitioning*: how much labeled carbon stays in fungal tissue versus appearing in the receiver's plant tissue. Sometimes the most honest result is that the fungus takes a great deal of carbon but little enters the neighboring plant, a finding that complicates any simple story of benevolent redistribution.

A small historical anecdote illustrates how much our confidence depends on the existence of the tag. Before stable isotopes became routine ecological tools, botanists sometimes tested belowground connections by girdling trees—removing a strip of bark to interrupt phloem flow—and watching downstream effects. If a girdled tree's roots declined and nearby seedlings suffered, one might argue for some shared dependency. But girdling is a blunt instrument. It alters hormones, water relations, and microbial communities; it changes shade; it can kill the tree. With isotopes, the experiment becomes gentler and more specific. Instead of breaking the system to see what collapses, you mark one stream of matter and watch where it goes. In a living forest, that distinction matters ethically as well as scientifically.

The label itself is also instructive about the forest's ordinary chemistry. Carbon-13 is rare in the atmosphere, and plants discriminate slightly against it during photosynthesis because the lighter carbon-12 diffuses and reacts a little faster. That natural discrimination creates characteristic δ^{13}C patterns that already tell ecologists about water-use efficiency and photosynthetic pathways. In other words, the forest has an isotopic texture even without experiments. A pulse labeling experiment superimposes a sharp, temporary pattern on that texture. The art is to make the pulse large enough to rise above background variability but not so large that it causes physiological disturbance. In practice, carbon-13 enrichment levels can be dramatic compared with natural variation, which is part of why the method works. The tree becomes, briefly, unmistakably different.

Placing the donor tree under a bag is also a reminder that the most sophisticated measurements can depend on simple field craft. A leaky seal at the stem can ruin the label. Condensation inside the bag can reduce light and alter stomatal opening. Wind can whip the chamber and stress the branch. Researchers learn to build chambers that are sturdy, transparent, and as non-invasive as possible. They time pulses to avoid overheating. They monitor carbon dioxide concentration so that the tree is not starved or flooded. The atomic breadcrumb trail begins, quite literally, with careful taping.

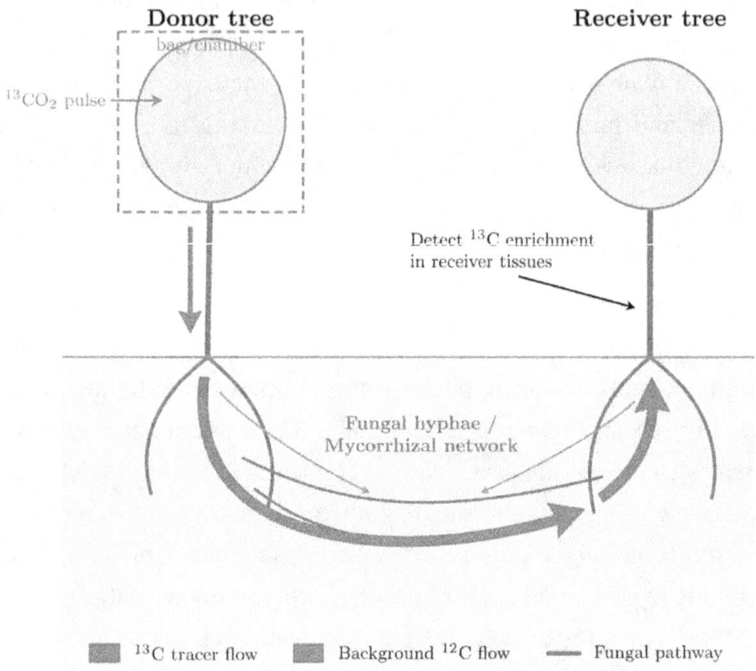

Schematic: a $^{13}CO_2$ pulse labels sugars in a donor tree; enriched carbon can move belowground into fungal hyphae and be detected later in a neighboring receiver tree.

The schematic hides a final subtlety: the measurement does not have to stop at "tree tissue." An isotope signal can be tracked into fungal-specific compounds, such as certain lipids that fungi make and plants largely do not. That matters because a mycorrhizal partnership is not a hollow pipe. Much of the carbon sent belowground may be used to build fungal biomass, or to fuel fungal metabolism, rather than to be exported to another plant. If the goal is to demonstrate a fungal route, finding enrichment in fungal biomarkers can be as compelling as finding it in a neighboring needle.

In the end, stable isotopes are powerful because they preserve the normal chemical identity of what is being tracked. A carbon atom in a sugar remains a carbon atom in a sugar. The forest continues its ordinary bargaining and spending. The only change is that some of the currency has been minted with a slightly heavier stamp, and careful accounting can reveal where it circulates.

6.2 Unmasking the Invisible Architects

Walk through a forest after rain and the evidence seems decisive: mushrooms, brackets, puffballs, delicate shelves stacked like shingles on a dying log. For a long time, this was the fungal census. Naturalists counted what they could see, pressed specimens, compared gills and spores and scents, and built a science of form. It was good science, often beautiful, and it gave fungi their public face.

It also gave a distorted map of the underground.

The familiar mushroom is not the organism in the way an apple is the organism. It is closer to an apple *tree*'s fruit: a temporary reproductive structure pushed up when conditions are right, then gone. The bulk of the fungus is the mycelium, a web of microscopic threads (hyphae) that lives inside soil, leaf litter, rotting wood, or root tips. Mycelium can persist for years without producing anything a walker would notice. Some fungi fruit unpredictably, or only briefly, or not at all in a given season. Others make fruiting bodies that are tiny, hidden, or easily misidentified. Even when a mushroom is visible, it is not always clear which tree roots the fungus is partnered with, or how far its mycelium extends, or whether the same fungal individual is connecting multiple hosts.

The problem is not merely that the eye misses things. The eye selects for drama. A mushroom survey records the species most willing to announce itself. The quiet majority remains uncounted, and in a networked system, missing nodes can change the story entirely.

Genetic identification changed the rules by making the underground legible. Instead of asking, *What fungi can we see?* ecologists began asking, *What fungi are actually here?* The difference is like the shift from identifying birds by feather collections to identifying them by the calls recorded through a wall. It replaces a census of appearances with a census of traces.

The basic idea is simple enough to explain at a dinner party. All living cells contain DNA, a chemical text written in four letters. Different species have different sequences. If a pinch of soil contains fungal threads, fragments of fungal cells, and fungal spores, then it contains fungal DNA. If you can extract that DNA and read a diagnostic piece of it, you can identify who was present in that soil even if no mushroom ever appeared. The trick is choosing the right piece of DNA: one that varies enough to distinguish species, but is conserved enough to be amplified reliably across a wide range of fungi.

For fungi, the workhorse is a region called ITS, the internal transcribed spacer. It sits in the ribosomal DNA, a part of the genome involved in making ribosomes, the cell's protein factories. Ribosomal DNA is present in many copies, which helps when you are trying to detect small amounts of material. The ITS region tends to differ among species, often enough to act like a barcode. A barcode does not need to tell you everything about an organism; it only needs to tell you which organism it is, the way supermarket stripes identify a product without describing its ingredients.

Using ITS as a fungal barcode did not instantly solve identification. A barcode is only as good as the catalog that links patterns to names. Early sequence databases were patchy and occasionally wrong, a reflection of the fact that fungal taxonomy itself was, and remains, a moving target. A single species might have multiple names, or multiple species might have been lumped under one name. Many sequences came from unidentified samples. Yet even imperfect genetic surveys revealed something that mushroom counts could not: soil communities were staggeringly diverse, and their composition changed over short distances, across seasons, and with host identity.

A concrete moment of realization came as researchers began sampling roots directly. Mycorrhizal fungi do not merely float through soil; they form intimate structures on or within roots. If you take a fine root tip and look under a microscope, you might see a fungal sheath (in ectomycorrhiza) or branching arbuscules inside the root's cells (in arbuscular mycorrhiza). Those structures are unmistakably fungal, but they rarely tell you *which* fungus. Under a microscope, many fungi look like pale threads.

So ecologists began to cut tiny root tips, extract DNA, amplify the ITS region using PCR (polymerase chain reaction), and sequence it. PCR is a molecular photocopier: it uses short primers that bind to conserved parts of DNA and then copies the region between them over and over until there is enough to read. Suddenly, each root tip could be tagged with a fungal identity. A tree that appeared to have a single kind of mycorrhiza under the microscope might host dozens of fungal partners across its root system. Two neighboring trees might share some fungi and differ in others. A single fungus might appear on many hosts. Networks that had been drawn in imagination could be checked against the presence or absence of shared genetic signatures.

Then came the next escalation: metabarcoding. Instead of sequencing one root tip at a time, metabarcoding sequences many DNA fragments from a mixed sample at once. A scoop of soil becomes a library of genetic snippets. High-throughput sequencers read millions of short sequences in parallel. The output is not a neat list but a flood: hundreds to thousands of distinct sequence variants in a single sample, each representing a taxon, sometimes a known species and often something unnamed.

This is where the romance of technological omniscience can mislead. Metabarcoding does not hand you a perfect map of the fungal web. It hands you evidence, and evidence needs interpretation. A soil sample is not a terrarium. DNA can come from living mycelium, dormant spores, recently dead tissue, or fragments transported by water. Different species' DNA amplifies with different efficiency depending on primer match and copy number. A rare sequence might represent a rare organism, or a common organism that amplifies poorly. A common sequence might represent abundant mycelium, or simply a species with many copies of ribosomal DNA. The genetic data tell you who left traces, not necessarily who is active or influential.

Still, the method is transformative because it gives fungi a kind of presence even in their quietest state. It makes it possible to ask: are the same fungal species found under different trees? Does a disturbance alter the community? Does drought simplify the network? Does a seedling recruit the same fungal partners as an adult tree? Those questions are not decorative. In forests, fungal partnerships influence how trees acquire nitrogen and phosphorus, how they tolerate stress, and how carbon enters soil. If the fungal community shifts, the forest's metabolism shifts with it.

One of the best historical anecdotes in this story is also a warning about appearances. In 1993, a set of honey fungus (Armillaria) samples from Oregon helped reveal what is often described as one of the largest living organisms on Earth: a single fungal individual spanning many hectares, spreading through roots and soil. The organism was not discovered by a spectacular mushroom display; it was inferred by genetic identity across distant samples. Whether one emphasizes its size or the later refinements to how such measurements are made, the cultural point remains: mycelium can be vast, coherent, and hidden. Visible fruiting bodies are the occasional flags of an organism whose borders are otherwise invisible.

As sequencing became cheaper and more standardized, the community problem shifted from *Can we read the DNA?* to *Can we interpret it consistently?* This is where curated reference databases and careful bioinformatic methods became as important as field work.

For fungi, one of the most widely used reference resources is UNITE, a curated database focused on fungal ITS sequences. Its value is not only that it contains many sequences, but that it tries to organize them into stable "species hypotheses": clusters of similar ITS sequences that likely correspond to species-level groups, even when those species have no formal Latin name. This matters because environmental sequencing uncovers an enormous number of fungi that have never been described morphologically. Without a way to label them, they remain a fog of anonymous reads. A stable hypothesis identifier is a way to say, *this same taxon appears here and there*, even if taxonomy has not caught up.

Taxonomy, in fungi, often lags because the organisms do not cooperate with the older rules. Many species cannot be cultured easily. Many fruit rarely. Many are cryptic, meaning they look similar while being genetically distinct. Some fungal lineages have tangled histories of naming. UNITE, by acknowledging uncertainty and keeping track of alternative classifications, treats identification as probabilistic rather than absolute. That attitude is not academic fussiness. It is the difference between a map that invites overconfidence and a map that admits where the coastline is still unmapped.

Then there is the problem of distinguishing genuine biological variation from sequencing noise. High-throughput sequencing generates errors: a base misread here, an insertion there. Early metabarcoding studies often clustered sequences into "operational taxonomic units" (OTUs) based on a similarity threshold (say, 97%). This was practical but crude. Two distinct species might be lumped together; a single species might be split into several OTUs because of errors. That matters acutely when you are trying to infer shared fungi between trees. If your method artificially splits one fungus into many, you will underestimate sharing. If it lumps several fungi together, you will overestimate sharing.

Methods like DADA2 improved this by modeling sequencing errors and inferring exact amplicon sequence variants (ASVs). The idea is to treat the sequencing run as a noisy observation of an underlying set of true sequences, then use statistics to reconstruct those true sequences. Instead of deciding that anything within 3% is "the same," ASVs aim to be exact sequence types. This can increase resolution and reduce spurious diversity. It also raises the stakes for good laboratory practice, because contamination and index hopping can appear as real ASVs if one is not careful. The reward is sharper inference: when the same ASV appears on different root tips, it is stronger evidence of shared taxa than a fuzzy OTU match.

Once taxa are identified, another temptation arises: to translate names into jobs. Ecologists want to know not only *who* is there, but *what* they do. Some fungi are saprotrophs that decompose dead organic matter. Some are pathogens. Some are mycorrhizal mutualists. Yet a sequence does not wear a badge. Tools like FUNGuild attempt to assign ecological roles based on taxonomy: if a sequence matches a genus known to be ectomycorrhizal, it may be assigned that guild. This is useful and often the only practical path to functional interpretation at scale. It is also inherently uncertain. Many taxa contain species with different lifestyles. Some taxa are poorly studied. Environmental sequences can match only to high taxonomic levels, where guild assignment becomes vague. A responsible analysis therefore carries the uncertainty forward: many reads remain unassigned, and many assignments are tentative.

Here is where the genetic approach becomes most exciting for a reader interested in networks rather than in lists. Metabarcoding can reveal overlap patterns: which fungal taxa tend to co-occur with certain hosts, which appear broadly, and which are finicky. In some forests, ectomycorrhizal communities are dominated by a few common taxa that appear on many trees, with a long tail of rare taxa. In others, the community is more evenly spread. In some cases, seedlings host a subset of the community found on adult trees, suggesting either selective recruitment or simple opportunity. In other cases, seedlings harbor distinctive fungi, perhaps because their roots occupy different microsites or because their chemistry differs.

A single soil core can also contain multiple worlds stacked together. The litter layer is rich in decomposers; the mineral soil may favor different fungi; the rhizosphere (the soil immediately around roots) is a chemical neighborhood shaped by root exudates. Sampling depth and compartment matter. A mushroom survey collapses these layers into a single impression. A genetic survey, if designed well, can separate them and reveal that fungi partition the habitat in ways the surface cannot show.

There is an emotional element to this, too, and it is not sentimental. Forests are often discussed in terms of the charismatic: the age of trees, the height of canopies, the return of wolves. The genetic portrait forces attention to the small and numerous. A handful of soil can contain a crowd as complex as a city. The invisible architects are not merely present; they are structured. They have preferences, competitors, and histories.

At the same time, genetics can puncture comforting narratives. A popular image of the "wood wide web" imagines a clean, benevolent network with a few dominant fungal conduits linking all trees into a cooperative whole. Genetic maps rarely look so tidy. They show overlap, yes, but also patchiness. They show that two trees a meter apart can host surprisingly different communities. They show that a single root system can carry a mosaic of fungal partners. They show that what looks like a single network from a distance may in fact be many partially connected subnetworks, with boundaries that shift as roots grow, die, and regrow.

The most honest use of genetic identification is therefore not as a proclamation of universal connectedness, but as a way of making hypotheses testable. If two trees share many fungal taxa in their roots, it becomes plausible that hyphal connections might exist between them, though it is not guaranteed. If they share none, then direct connection by a common mycorrhizal fungus is unlikely, though they may still influence one another through the soil community in other ways. If a disturbance reduces diversity and increases dominance by a few taxa, one can ask whether that changes nutrient cycling or resilience. If drought favors fungi with certain traits, one can ask what that means for tree survival.

The genetic approach also encourages a different kind of patience. A mushroom appears when the fungus chooses to reproduce; DNA appears whenever the fungus lives, dies, or passes through. Sequencing captures an average of presence over some window of time, an integrated record of biological activity and debris. This can be frustrating when one wants a crisp picture of a moment. It can also be valuable, because forests are not momentary systems. They are systems in which slow processes matter: the gradual accumulation of organic matter, the slow turnover of roots, the long-term persistence of fungal individuals.

A forest manager trying to restore a degraded site may care about whether ectomycorrhizal partners are present for certain tree species, or whether pathogens dominate, or whether fungal diversity rebounds after compaction. A conservation biologist may care about whether old-growth stands harbor distinct fungal lineages not found in managed forests. A climate scientist may care about whether shifts in fungal communities alter soil carbon storage. In each case, the ability to identify fungi without waiting for mushrooms turns an invisible process into something measurable.

All of this begins with a small act that would have baffled earlier naturalists: taking a smear of soil, dissolving the cells into a chemical soup, and reading a piece of text from the wreckage. It feels destructive, almost rude. Yet it is also a way of listening. The mycelium writes itself into the soil as DNA, and the sequencer translates that writing into names, hypotheses, and, with luck, a clearer sense of the living geometry beneath our feet.

6.3 The Proof of Exclusion

The trouble with invisible connections is that they are easy to believe in and hard to prove. A thread of fungal hypha is thinner than a human hair by an order of magnitude; it can pass through pores in soil that water flows through but roots cannot. That physical delicacy invites a particular kind of scientific anxiety: if something moves from one plant to another, how do you know it went through the fungal web rather than around it?

Stable isotopes can mark a molecule with an unmistakable signature. DNA sequencing can tell you which fungal names are present near which roots. Neither, on its own, guarantees the route. Carbon might move as carbon dioxide through air pockets, or as dissolved compounds through water films, or by being eaten and excreted by microbes, or by simple diffusion. Two adjacent plants can exchange resources without fungi at all, through root contact, root grafts, or the ordinary chemistry of soil. A beautiful isotopic enrichment curve can therefore be a kind of mirage: a true signal of movement, attached to the wrong mechanism.

Ecologists call the corrective move a *proof of exclusion.* It is the logic used whenever the world offers too many plausible explanations. If you cannot directly watch the process, you remove one pathway while leaving the others as intact as possible, and you see what changes. The art lies in the phrase "as intact as possible." Forests are not circuits in a lab bench. Every barrier you insert disturbs something. The goal is not perfection; it is a disturbance that is precise enough to be interpretable.

The most widely used barriers are meshes, chosen not for their symbolism but for their pore size. Roots and fungi are built on different scales. A fine root tip, even a delicate one, is measured in tenths of millimetres. A fungal hypha is commonly measured in micrometres. A mesh with pores around a few tens of micrometres can allow hyphae to pass while physically blocking root tips. A much finer mesh, with pores around half a micrometre, can block hyphae while still allowing water, dissolved nutrients, and many bacteria to move through. The numbers vary with the fungus and with the practicalities of manufacturing, but the principle is robust: pores can act as a gatekeeper for biology.

It is a deeply satisfying idea because it uses the simplest physics. You do not need to know which genes the fungus has or what enzymes it secretes. You only need to know that a thread cannot pass through a hole smaller than itself.

Imagine a seedling planted inside a cylinder of mesh in the soil. Outside the cylinder are the roots of an older tree and a living mat of mycorrhizal hyphae. If the mesh pores are wide enough for hyphae, fungal threads can grow through the mesh and colonize the seedling's roots. If the pores are too small, the fungus is kept out, even though water and dissolved nutrients can still seep in and out. The seedling experiences the same soil moisture and much of the same soil chemistry, but it loses physical access to the fungal network. If the seedling's growth, nutrient uptake, or isotopic enrichment changes dramatically between these two conditions, the fungal route becomes a far more plausible explanation.

This kind of design has been refined into a family of experiments with small differences and large consequences. Sometimes the mesh cylinder is called an in-growth core, because it allows hyphae to grow into a defined volume of soil. Sometimes researchers install paired cores: one with hyphal access and one intended to exclude hyphae. Sometimes they add a third condition that excludes both roots and hyphae, using a solid barrier, to estimate purely chemical diffusion. Each refinement tries to address a skeptical question before it is asked.

Skepticism is not a mood; it is a method. Consider what happens if you only block roots but allow fungi. The seedling can still take up nutrients directly through its own roots from the soil water that passes through the mesh. If it shows increased uptake of a labeled nutrient compared to a seedling with hyphae excluded, that difference suggests hyphal delivery. But it might also reflect indirect effects. Hyphae can change soil microbial communities; they can alter nutrient mineralization; they can change pH in microscopic zones. A barrier that admits hyphae therefore admits the fungus's influence as a living organism, not merely as a pipe. This is often exactly the point, but it complicates interpretation if the claim is narrow (for example, "phosphorus moved through hyphae") rather than broad ("fungi altered nutrient availability").

That is why some of the most persuasive proof-of-exclusion designs do not merely block or allow hyphae. They actively *sever* them.

One clever approach uses rotation. A mesh core is installed so that hyphae can grow into it. Then, at regular intervals, the core is rotated slightly, cutting the hyphae that cross the boundary while leaving the soil and the plant inside largely intact. The difference between an unrotated core (hyphae continuously connected) and a rotated core (hyphae repeatedly cut) can be striking. If a labeled nutrient appears in the plant only when hyphae are allowed to remain connected, and not when they are repeatedly severed, the inference becomes stronger: whatever is being delivered depends on intact hyphal continuity, not merely on the presence of fungal biomass somewhere in the neighborhood.

This idea has a pleasing physical bluntness. It does not require an antibody or a gene knockout. It simply recognizes that networks are defined by connection, and it tests what happens when connection is lost.

A concrete example comes from early work in grassland systems, where researchers used radioisotopes such as phosphorus-33 to track nutrient movement. Phosphorus is famously immobile in many soils; it binds to minerals and becomes difficult for roots to access. Mycorrhizal fungi, with their fine hyphae, can access phosphorus from tiny pockets and transport it back to roots. In in-growth core experiments, plants with hyphal access often take up orders of magnitude more labeled phosphorus than plants from which hyphae are excluded or severed. The appeal of these studies is not that they are dramatic, though they are. It is that they isolate a mechanism that makes sense in the real world: the geometry of exploration matters when a nutrient is patchy and sticky.

Geometry also explains why exclusion experiments need careful verification. A mesh is only as good as its seams. Soil shrinks and swells with moisture. Roots probe, press, and sometimes exploit gaps. Hyphae can be astonishingly opportunistic, slipping through tiny defects. A core that was meant to exclude hyphae can become colonized anyway, producing a false negative ("no difference between treatments") or a false positive ("hyphae excluded but label still moved, so hyphae must not matter") depending on how the failure is interpreted.

Modern studies often verify exclusion success using molecular tools such as quantitative PCR (qPCR), which can estimate the amount of fungal DNA in a sample. If an exclusion core is meant to stay hypha-free, qPCR can reveal whether fungal DNA is nonetheless present. Researchers can also examine root colonization directly or measure fungal biomarkers. These checks can feel like tedious housekeeping, but they are the difference between a persuasive design and a persuasive-looking design.

There is also a more subtle failure mode: confusing fungal enrichment with plant enrichment. When a labeled carbon pulse is given to a donor plant, the label can move into fungal tissues in a receiver compartment even if little or none is incorporated into the receiver plant itself. This is not a trivial distinction. A mycorrhizal fungus might be connected to two plants and still allocate most incoming carbon to its own growth or respiration rather than passing it on. In a compartmented microcosm, you might detect the label in the receiver *compartment* and be tempted to say the receiver plant received carbon. Careful partitioning of samples—separating plant tissue from fungal tissue, and measuring each—prevents this kind of overinterpretation.

Such caution has become especially important because the idea of common mycorrhizal networks has acquired cultural momentum. The public story often runs ahead of the methods, turning a plausible mechanism into a moral fable about forest cooperation. The scientific debate has therefore sharpened around definitions and controls. What counts as a "network" in a field setting? Is it enough that two plants host the same fungal species, or must the same fungal individual physically connect them? Does isotopic enrichment in a neighbor imply active allocation, passive leakage, or microbial recycling? Is a small transfer ecologically meaningful, or merely detectable?

Proof-of-exclusion experiments do not answer every question, but they do force the conversation to stay close to mechanism. They encourage conservative claims: *when hyphae are physically blocked, the effect disappears; when hyphae are allowed, the effect returns.* This is not a guarantee of benevolent behavior or intelligent coordination. It is a statement about the necessity of a pathway.

Even the word "barrier" can mislead if it suggests a wall. In soil, a barrier is a filter. Water carries dissolved ions through. Small organic molecules can diffuse. Bacteria can often pass, depending on pore size and water films. Fungal enzymes released outside the barrier can still act on organic matter and create soluble products that move inward. If an exclusion core shows reduced nutrient uptake, that difference might reflect the loss of hyphal transport, or the loss of enzyme delivery, or the loss of fungal influence on bacterial communities. Sometimes those are inseparable in practice, and sometimes that is acceptable because the question is ecological rather than anatomical: *does the presence of an intact mycorrhizal connection change outcomes for the plant?*

This is where experimental design becomes a kind of negotiation with complexity. A very tight mesh might exclude hyphae, but it might also change water flow, oxygen diffusion, and the physical structure of the soil. A looser mesh might admit hyphae but also admit fine roots from outside, creating the possibility of direct root competition or root grafts. Solid barriers can isolate diffusion, but they can also create unnatural moisture gradients. No design is neutral. Good studies therefore rely on comparisons among multiple barrier types, on replication across many plants and sites, and on measurements that track not just the plant but the soil conditions inside and outside the barriers.

In forests, another complication is time. A fungal network is not a static cable. Hyphae grow, die, and regrow. Roots turn over. A barrier installed in spring may become irrelevant by late summer if roots proliferate around it or if hyphae find a way through. Some experiments are therefore short-term by necessity; they capture a snapshot of connectivity under controlled conditions. Long-term exclusion is harder, and when achieved it may introduce its own artifacts. The most careful interpretations keep the timeframe explicit: a barrier experiment demonstrates what the system can do under certain constraints, not necessarily what it always does in every season.

A memorable, almost domestic image brings the logic home. Consider two neighboring households. If you suspect they share electricity through a hidden cable, you can measure their power usage and look for correlations. You can even add a tracer to one household's power supply and see whether the other's lights flicker in a characteristic way. But if you want to prove the cable exists, you eventually put a shovel in the ground and cut the line. In soil ecology, the shovel is replaced by a mesh and the cut is often a rotation, but the logic is the same: the most convincing evidence comes from a disappearance.

The deeper value of proof-of-exclusion is that it disciplines imagination without extinguishing it. Fungal hyphae are real conduits, but they are also living organisms with their own economics. They transport phosphorus and nitrogen to plants, but they also demand carbon. They connect roots, but they also compartmentalize. They can create pathways for resource movement, but they can also create competition among plants linked to the same fungal partner. Exclusion experiments, when done carefully, do not romanticize the network. They reveal its necessity under certain conditions and its irrelevance under others.

In some settings, blocking hyphae produces a dramatic effect on seedling nutrition and growth. In others, the effect is subtle or absent. Those "absent" results are not failures. They are part of a mature picture: networks are not universal cables delivering constant aid. They are contingent structures whose influence depends on soil fertility, moisture, host identity, fungal community composition, and the broader microbial context. A fungal web can be a lifeline in one patch of soil and a minor player in another only a few metres away.

The most persuasive experiments therefore combine the atomic breadcrumb with the barrier. A labeled nutrient or carbon pulse provides a signature; the barrier tests the route. If the label appears in a seedling only when hyphae can physically cross, and not when they are excluded or repeatedly severed, then the ghostly signal gains a body. It becomes a mechanism that can be argued with, refined, and, when necessary, doubted for specific reasons rather than dismissed as folklore.

Under a forest floor, proof rarely arrives as a single decisive observation. It arrives as a pattern of constraints: threads that cannot pass through a pore; labels that disappear when a connection is cut; fungal DNA that refuses to stay out unless the mesh is intact; a seedling that thrives only when the microscopic bridge is allowed to form. In that accumulation of careful negatives, the underground network becomes something sturdier than an idea.

Chapter 7

Strength in Fragile Connections

Forests withstand storms and droughts not through the brute strength of individual trees, but through the statistical power of their underground connections. We examine the architecture of this biological insurance, revealing how redundancy creates robustness and pinpointing the terrifying mathematical tipping point where a coherent ecosystem suddenly shatters into fragments.

7.1 The Architecture of Insurance

A forest is full of bargains struck in the dark. A tree sends sugars downward into its roots—the hard-won product of sunlight—and in return receives help gathering what the air cannot provide: phosphorus locked in minerals, nitrogen scattered thinly through soil, water held in capillaries too narrow for roots to drink from directly. The helpers are fungi, and their generosity is not charity. They are trading partners. They are also, from the tree's point of view, a form of insurance.

Insurance is not a warm, fuzzy idea. It is a cold one. It assumes that bad things happen and that they happen unevenly. One house on a street burns, not all of them. One crop fails in one valley, not in every valley. A drought arrives, but its harshness comes in patches: this south-facing slope bakes; that shaded hollow stays damp. If life had to choose between stability and flexibility, it would choose flexibility every time—but flexibility is costly unless it is organized.

Biological insurance begins with a simple refusal: trees rarely commit to a single fungal partner. Beneath even a solitary trunk there is a crowd. Roots are colonized by multiple species, sometimes by dozens, forming a shifting portfolio of relationships. The portfolio is not an ornament of diversity. It is an architecture: redundancy that prevents a local failure from becoming a systemic one.

Redundancy can sound like waste. Engineers spend careers removing it. Why build two bridges when one will do? Why keep multiple suppliers when one is cheaper? Yet the world that trees inhabit is not a controlled factory floor. It is a place where a minor heatwave can unravel delicate hyphae, where a soil pocket can turn acidic after a rain, where a beetle outbreak changes the chemistry of root exudates, and where a fire can sterilize one centimeter of soil while leaving the next centimeter alive. In such a world, a single point of dependence is not efficiency; it is fragility.

The fungal body, most of the time, is not a mushroom. It is a web of threads—hyphae—that extend outward from a colonized root, probing soil pores, wrapping around grains of sand, fusing into cords that can run like cables through the litter. Those threads are thin enough to slip into spaces a root hair cannot enter. They are also disposable. Hyphae can die back quickly when conditions sour, and the fungus can regrow when conditions improve. The tree, watching this from the inside, cannot afford to bet its nutrition on one fragile extension cord. It needs backups.

One way redundancy appears is through overlap of function. Different fungi can perform similar services. Several partners can deliver phosphorus; several can liberate nitrogen from organic matter; several can increase drought tolerance by extending the effective reach of roots. If one species declines, another can take over *enough* of the work to keep the tree alive. The goal is not perfection. It is continuity.

But redundancy is not only about having multiple partners who can do the same job. It is also about having partners who do the job differently. Some fungi send out dense, fine foraging networks that harvest nutrients close to the root surface; others build thicker cords that specialize in long-distance transport. Some tolerate drought; others flourish in cool, moist soils. Some are rapid colonizers of disturbed sites; others are slow and persistent, holding territory year after year. A diverse fungal portfolio gives a tree not just a spare tire, but a set of tires suited to different roads.

There is an old lesson from finance here. A portfolio does not protect you because it contains many assets; it protects you because those assets respond differently to shocks. If everything in your portfolio rises and falls together, you are exposed. Forest fungi are similar. Redundancy pays off when partners are *not* perfectly correlated in their failures.

In the late twentieth century, ecologists began to realize that these underground arrangements were not simply many separate partnerships repeated across a landscape. They could form networks in which multiple trees are connected through shared fungal individuals. That possibility changed the question from "which fungus belongs to which root?" to "what kind of network is this, and how resilient is it?"

One of the most vivid concrete examples comes from Douglas-fir forests in western North America, where researchers mapped ectomycorrhizal fungi of the genus *Rhizopogon*. Ectomycorrhizal fungi form a sheath around fine roots and send hyphae into the soil; they are particularly common in many temperate and boreal forests. By identifying fungal *genets*—genetic individuals that can sprawl underground and colonize multiple roots—the researchers could draw a literal network: trees as nodes, and shared fungal individuals as connections. What emerged was not a tidy set of one-to-one partnerships. It was a tangle of overlapping links, with some fungal individuals connecting multiple trees, sometimes across different age cohorts. A young tree and an older neighbor might not be exchanging friendly greetings; they might simply be plugged into the same living infrastructure.

That mapping effort matters because it gives the network metaphor a backbone. It is tempting to talk about "the wood-wide web" as a poetic flourish. Yet in those Douglas-fir stands, the "web" was not just inferred from patterns of growth. It was reconstructed from genetic evidence of shared fungal bodies. The existence of that structure does not, by itself, guarantee dramatic sharing of resources in all directions at all times. The strongest claims—that mature trees preferentially feed their seedlings, or that forests behave like cooperative superorganisms—remain hotly debated, and critics have rightly pointed out how easy it is to overinterpret limited experiments. But even without romance, the network has a sober implication: if a single fungal individual links several roots and several trees, it can act as a conduit when conditions make some patches of soil richer than others. It can also act as a failure point when that individual is stressed.

Here redundancy becomes not a luxury but a necessity. A star-shaped network—one central hub connected to many leaves—is efficient: the hub can coordinate traffic; materials can move with few steps. But it is brittle. Lose the hub, and the leaves are isolated. A mesh network—many nodes connected with multiple overlapping paths—can look chaotic, even "wasteful." Yet it is robust. Block one route, and traffic detours.

Forests contain both kinds of patterns, at different scales. At one moment and in one patch of soil, a particularly successful fungal individual may behave like a hub. At another scale, the community as a whole may form a mesh because trees host multiple fungi and fungi colonize multiple trees. The insurance comes from the mesh-like redundancy: the presence of alternative paths.

This is not just abstract network talk. Consider what happens during a dry summer. Water becomes the limiting currency. Some fungi are more drought-tolerant than others; their hyphae resist desiccation, or they retreat to deeper, moister layers. A drought-sensitive fungal partner may reduce its activity precisely when the tree most needs help. If the tree has only that partner, it is stranded. If the tree has additional partners—perhaps one with deeper cords, perhaps another whose hyphae persist longer in dry pores—the tree has options. The forest, too, has options: more links remain functional, even as others fail.

There is a related idea from plant physiology that makes the image vivid. Some plants can perform "hydraulic lift": they draw water from deeper layers at night and release some of it into shallower, drier soil through their roots. In principle, a mycorrhizal network could provide pathways that redistribute some of that water, moving it toward roots that did not themselves reach the deep reservoir. Whether this happens widely and at ecologically decisive scales is still an active research question, but the possibility highlights an important point: the underground network can transmit not just nutrients but buffering capacity. The *potential* for redistribution is itself a form of insurance. Even if the average flow is small, the timing matters. A small amount of water at the right hour can keep fine roots alive. A small amount of phosphorus in a short window can determine whether a seedling sets enough needles to survive winter.

Insurance also operates in the microbial realm by dampening the consequences of specialization. Some fungi are generalists: they can partner with many host species and tolerate a wide range of soils. Others are specialists: exquisite in a narrow niche but poor outside it. Specialists can be extraordinarily effective at extracting a particular nutrient or thriving under a particular chemistry. The catch is that the world rarely stays within the narrow band that specialists require.

A tree that hosts only specialists is like a traveler carrying three perfect keys for three perfect locks. It feels clever until the locks change. A tree that hosts a mixture of generalists and specialists is less optimized but far safer: when conditions shift, generalists keep the system running while specialists contribute their advantages when the niche reappears.

This mixture also shapes how failure propagates. If most roots rely on a single specialist and that specialist is knocked back—by heat, by a pathogen, by a sudden change in soil pH—the effect is synchronized. Many roots lose service at once. If roots are partitioned among multiple partners, failures are desynchronized. Some roots become less supplied while others remain serviced. In systems theory, desynchronization is a powerful stabilizer. It prevents the whole system from tipping at once.

There is a historical echo here, not from forests but from human agriculture. The Irish potato famine of the 1840s remains the textbook case of what happens when an entire food system rests on low diversity. A pathogen, *Phytophthora infestans*, exploited the uniformity of potato varieties, and the consequences cascaded through society. Forests are not immune to analogous vulnerabilities. A monoculture plantation, even if it looks green and healthy, can have belowground uniformity that leaves it exposed. Diverse fungal partnerships are one of the ways natural forests avoid the biological equivalent of putting every calorie into a single crop.

Redundancy, though, is not merely a passive collection of partners. It is organized by selection pressures acting on both sides of the symbiosis. Trees "choose" in the only way a tree can: by allocating carbon. A fungal partner that delivers more limiting nutrients can be rewarded with more sugars. A partner that delivers little may be starved. This is not conscious bargaining, but it is a negotiation shaped by evolutionary feedback. Fungi, likewise, allocate their growth toward roots that supply them well. The result is a constantly updated map of investment.

That map can create the very mesh-like redundancy that makes the system resilient. A tree does not need to maintain equal relationships with all partners. It needs to avoid being trapped. It needs a set of viable alternatives in the soil—partners that are present, connected, and ready to expand when a competitor falters.

One concrete way to picture this is to imagine the root zone as a city. Each fungal species is a transportation company with its own fleet: bicycles, buses, freight trains. The tree is a shipper that wants reliable delivery of scarce goods. If all deliveries depend on one company and one bridge, a single accident blocks the city. If several companies operate, using different routes and vehicles, then a flood that washes out a bridge does not halt commerce; it reroutes it.

The redundancy is especially valuable because fungal networks can be locally damaged in extremely local ways. Hyphae can be severed by soil animals; a pocket can dry out; a toxin can accumulate near a decomposing log; a root can die back after herbivory. Each event is a small tear. What matters is whether the tear isolates the whole system or whether traffic finds another way.

This is why the simplest network diagrams are so useful: they show the difference between connectivity and dependence at a glance. A star topology looks connected, but it depends on the hub. A mesh topology may look less elegant, but it tolerates breakage.

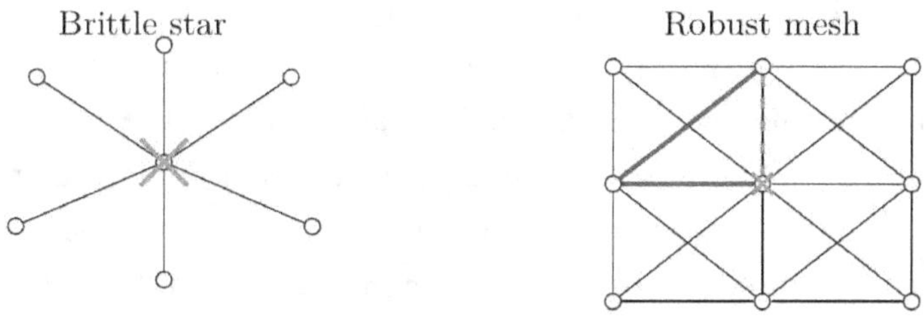

Single failure disconnects many

Failure reroutes flow

The point of such a sketch is not to claim that real fungal networks look like tidy lattices. They do not. They are lopsided, contingent, and constantly remodeled. The sketch simply captures the logic of insurance: alternative routes reduce the consequences of any one break.

Of course, redundancy is not free. Every fungal partner expects payment in carbon. For a tree, carbon is time: each gram of sugar is a fraction of a day spent photosynthesizing. In a bright summer, a mature tree can afford more partners than a shaded seedling can. This introduces a humanly recognizable tension. We buy more insurance when we can afford it, and we drop coverage when times are tight. A struggling seedling might not be able to support a wide portfolio. That is one reason why establishment is the perilous moment in a forest life cycle, and why access to existing fungal networks in the soil can matter so much. If a seedling germinates into a patch where compatible fungi are already active, it can plug into a preexisting service network rather than paying the full cost of building one from scratch.

The idea of a "portfolio" also helps temper a common misunderstanding. When people hear that trees can be linked underground, they sometimes imagine a benevolent community in which everyone shares. But redundancy is compatible with competition. A network can buffer individuals without being altruistic. A power grid supplies electricity to rival businesses. A road system carries goods for competing stores. Connectivity allows flows; it does not dictate motives. In forests, carbon, nitrogen, phosphorus, and water tend to move along gradients of supply and demand—from where they are more available to where they are more needed—and those gradients shift hour by hour. A shaded tree can become a carbon sink; a sunlit neighbor can become a source. In one season a root patch might be phosphorus-rich; in another it might be depleted. The network's value, as insurance, is that it keeps the possibility of exchange open when the environment makes exchange useful.

One can see this in the way networks respond to partial loss. Suppose a particular fungal species is hit by a heatwave and its hyphae retract. If other fungi remain, and if the tree maintains at least some colonization with them, nutrient flow does not stop. It changes. The tree might receive less of one nutrient and more of another; the spatial pattern of uptake might shift; growth might slow. Yet the tree remains in the game. The forest avoids the sudden transformation from "mostly functioning" to "not functioning at all." That avoidance—the smoothing of shocks—is what insurance is for.

Redundancy also provides insurance at the scale of the forest, not just the individual tree. Different tree species often associate with different fungal communities. A mixed forest can therefore host a richer fungal diversity than a single-species stand. That fungal diversity can, in turn, influence which tree species recruit successfully after disturbance. The relationship is reciprocal and a little unsettling: the future composition of the forest can be shaped by the invisible composition of its fungal partners, while the fungal community is sustained by the living plants that feed it. Insurance, in this sense, is mutual: a diverse forest supports a diverse fungal community, and a diverse fungal community supports the forest's capacity to persist through variable conditions.

There is a final twist. Redundancy does not only mean multiple routes for the same kind of resource. It can mean multiple *forms* of storage and persistence. Some fungi persist as long-lived mycelial systems; others as spores in the soil waiting for the right host; others as resistant structures that endure drought or frost. A forest that contains a variety of fungal life histories is not just robust to a broken link; it is robust to time. When a disturbance temporarily wipes out active hyphae near the surface, deeper cords and dormant propagules can repopulate the space. The insurance policy is written not only in space, as alternative pathways, but in time, as alternative ways of surviving interruption.

To walk through a forest is to see trunks and leaves and light. To live as a tree is to experience risk in the soil: a patch of drought here, a surge of pathogens there, a shortage of phosphorus in one pocket and abundance in another. The architecture that keeps those local troubles from becoming fatal is not a single miraculous thread connecting everything. It is a quilt of overlapping partnerships—redundant, uneven, continuously renegotiated—that makes it possible for a living organism rooted in place to endure a world that will not hold still.

7.2 The Threshold of Collapse

Forests are good at looking fine while quietly becoming less connected. A canopy can stay green, trunks can keep thickening, and seedlings can keep appearing at the edges of clearings even as the underground plumbing is being simplified, thinned, and broken into smaller pieces. Part of the reason is psychological: aboveground, continuity is easy to see. Belowground, continuity is something you infer.

Network science offers a language for this kind of hidden danger, and one of its most unsettling ideas is a threshold that arrives without warning. The formal term is *percolation*, a word borrowed from the everyday act of water finding a path through coffee grounds. Mathematicians and physicists began studying percolation seriously in the mid-twentieth century—Broadbent and Hammersley (1957) gave it its modern footing—as a way to ask a deceptively simple question: if you randomly block some fraction of pathways in a large grid, when does flow across the system become impossible?

It turns out that the answer is rarely "gradually." In many networks there is a tipping point: you can remove connection after connection and still find a route across, and then, after one more removal, the system abruptly shatters into islands. The technical phrase for the sprawling cluster that spans much of the network is the *giant component*—the largest set of nodes that can all reach one another through some chain of links. The crucial point is that the giant component can persist stubbornly even as damage accumulates, and then disappear suddenly when a critical fraction of links is lost.

A forest's mycorrhizal web is not a neat grid, and the links are not simple wires. Yet the percolation metaphor is more than decorative. Roots and fungal hyphae form a transport network for phosphorus, nitrogen, water, and sometimes carbon. That network is constantly remodeled: hyphae die back; roots turn over; fungi compete; soil cracks in drought; pores flood in rain. Disturbance is not an exception but the normal condition. What matters is whether the remaining structure is still one connected fabric or many separate scraps.

Percolation theory gives a way to explain a feeling many ecologists have had in the field: systems can absorb a surprising amount of incremental loss. A forest can lose some large trees, some fungal diversity, some soil moisture, some coarse woody debris, and still function. Nutrient uptake measurements remain respectable; growth rates dip but do not crash; leaves look normal. The danger is that this apparent sturdiness can lull you into thinking there is no cliff edge.

To see why the cliff edge exists, imagine reducing a transportation network. Closing a few minor streets does not prevent a city from functioning. Drivers reroute. Close more streets and traffic becomes annoying but still possible. Then, at some point, a particular bridge closes and whole neighborhoods become unreachable, not because the bridge carried all the traffic, but because it was the last connector between two halves of the map. The city did not become worse one street at a time; it crossed a structural boundary.

The best-known modern parable of this sort is the Northeast blackout of August 2003. A series of failures—some technical, some managerial—cascaded through a power grid and left tens of millions without electricity. No single event was sufficient by itself. The grid endured earlier faults, and then it did not. Power grids are not forests, but they share the same uncomfortable property: connectivity can hide fragility. A system can continue to deliver, until it cannot.

In the fungal web, links are created when multiple roots are colonized by the same fungal individual or when fungi form a sufficiently dense set of connections that resources can, in principle, be routed through different pathways. Empirical mapping in Douglas-fir forests, using genetic identification of fungal individuals (genets), has made these links tangible rather than speculative. When researchers draw those networks, some trees emerge as highly connected and some fungal genets spread broadly. The picture suggests a web that can move resources around obstacles. It also suggests a web that could break apart if certain connections are lost.

Percolation helps frame the question more precisely: as links are removed—by drought, by soil compaction, by shifts in fungal community composition, by selective logging that changes root density and carbon supply—does the network shrink smoothly, or does it remain mostly connected until it crosses a threshold and fragments?

The idea of a threshold becomes clearest in an intentionally simplified model. Picture a lattice where each dot is a root tip or a colonized patch of soil and each line is a hyphal connection. Now start deleting lines. At first, deletion hardly matters: alternative routes remain. Later, routes become longer and more tenuous. Finally, there comes a point when the last set of cross-cutting links is severed and the network breaks into clusters that cannot reach one another. In percolation language, the giant component collapses.

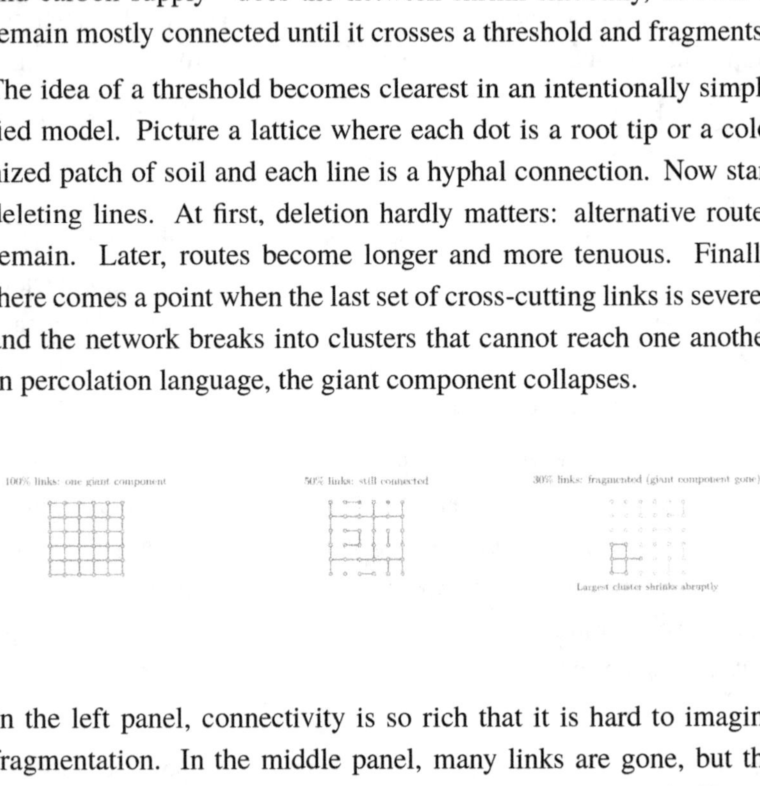

In the left panel, connectivity is so rich that it is hard to imagine fragmentation. In the middle panel, many links are gone, but the system still acts like one. In the right panel, one more round of losses does not merely thin the network; it changes its *kind*. Instead of one big system with some holes, you have many small systems with no way to pool resources. The phase transition is not a metaphor for drama. It is a geometric fact about routes.

This matters because ecological function often depends more on being part of a large connected component than on the absolute number of links. A small drop in connectivity can have a large effect if it disconnects a drought-stressed seedling from the only fungal pathways that reach a deeper, moister layer of soil, or if it isolates a patch of roots from fungal partners better at extracting phosphorus from a particular mineral pocket. Percolation language forces a change in intuition: the important variable is not "how many links remain?" but "does the network still span the relevant space?"

Once you start thinking this way, certain patterns in forests look newly ominous. Selective logging, for example, does not remove trees at random. It often removes the largest stems, the ones with the biggest crowns and the most extensive root systems. In a network, removing nodes at random can leave the giant component intact for a long time; removing hubs can break it quickly. This difference is one of the most robust findings in network theory, and it helps explain why a forest can be resilient to scattered small disturbances but vulnerable to disturbances that disproportionately target the best-connected individuals.

Belowground, the "hub" might be a large tree with many colonized root tips spread across a wide area. It might also be a particularly extensive fungal genet, one that occupies deep soil layers and links several trees. Either way, targeted loss changes topology. In a Douglas-fir system where *Rhizopogon* genets connect multiple cohorts, removing a subset of highly connected trees is not just removing photosynthetic capacity; it is tearing out junctions in the transport map.

Long-term studies of mycorrhizal networks in managed forests suggest that the picture can be even subtler. Whether a network appears fragile or robust can depend on what you count as a link. If you map only one fungal species, the network may look sparse and vulnerable; include a closely related species that occupies similar space, and suddenly you see redundancy that keeps the connected component larger. Van Dorp and Simard (2020) emphasized this kind of sensitivity when examining resilience decades after selective logging: conclusions about vulnerability can hinge on taxonomic coverage. This is not a minor technicality. It is a reminder that a forest's insurance policy is written in diversity, and that incomplete measurement can make a resilient system look brittle or a brittle one look resilient.

There is another network concept that deepens the percolation story: the k-core. A k-core is the subset of a network in which every node has at least k connections *within that subset*. The language is intimidating, but the idea is intuitive. A network can have a dense inner skeleton and a looser fringe. The fringe may fall away first with little visible effect. The inner skeleton is different. Once the skeleton is compromised, collapse can accelerate because removing a node does not just remove that node; it reduces the degree of its neighbors, which may then drop below the threshold and fall away too. This produces cascades that look, in ecological time, like sudden dieback.

Mutualistic networks in ecology—pollinators and plants, seed dispersers and fruiting trees—have been analyzed with k-core thinking, and it transfers naturally to mycorrhizal webs. If a small group of fungi and host plants form a densely connected core, they can stabilize each other. Yet that same structure can create a sharp tipping point. Peripheral partners can be lost quietly, but if the core is weakened (by repeated droughts, by soil disturbance, by chronic nitrogen deposition that shifts fungal communities), the remaining network can fall through its own internal trapdoor.

The trapdoor image can be misleading if it suggests a single critical moment you can predict precisely in a living forest. Real ecosystems do not behave like clean mathematical lattices. Links are not identical; some carry more nutrients than others. A connection that exists on paper may be physiologically inactive at a given time, because carbon supply has changed or because the fungus is reallocating its growth elsewhere. The network itself can change as species composition shifts: the loss of one fungus may encourage another to expand, partially compensating. These are good reasons to resist claiming that one can measure a forest, compute a threshold, and declare a date of collapse.

Yet percolation remains valuable because it teaches a disciplined form of fear: a respect for nonlinearity. It warns against trusting smooth trends. It also highlights why management actions that seem modest can have outsize effects when a system is already near a threshold. Compacting soil with heavy machinery, for instance, may not kill trees immediately but can reduce pore space and disrupt hyphal growth. Repeated prescribed burns may have beneficial effects aboveground but could, depending on intensity and frequency, simplify the fungal community in ways that erode redundancy. Chronic drought can make the loss of one fungal guild more likely and reduce the regrowth that would otherwise stitch the network back together.

A final complication is that thresholds can be history-dependent. Some ecosystems show *hysteresis*, a term borrowed from physics to describe systems that do not return along the same path by which they were pushed. It is easy to imagine how this could happen underground. If a connected mycorrhizal network fragments, seedlings may establish less successfully, reducing plant carbon inputs to fungi, which further reduces fungal abundance, which further limits seedling establishment. Even if rainfall returns to normal, the biological infrastructure that would have used that water has been diminished. In such a case, restoring the original conditions does not restore the original network quickly, because the network depended on its own continuity.

This is one reason the concept of a tipping point matters even when you cannot locate it precisely: it encourages humility about reversibility. The comforting assumption in many environmental discussions is that damage is linear and repair is linear. Percolation teaches the opposite. Connectivity can be lost quietly and regained slowly, especially when living partners must recolonize space, compete, and rebuild.

The most practical lesson is also the least glamorous. If you want a forest to endure a harsher climate—more drought, more heat, more intense fire seasons, more pest outbreaks—you do not only protect trunks and crowns. You protect the connective tissue that allows many small failures to remain small. That tissue includes diversity of fungal partners, continuity of soil structure, and the presence of well-connected trees and fungal individuals that act as bridges across space and time. When those bridges are gone, a forest can still look like a forest for a while. The change announces itself later, when the living network that made the forest coherent can no longer percolate.

7.3 Resurrection from the Soil

After a severe fire, the silence can feel accusatory. The canopy that once softened wind and light is gone, the air smells of mineral ash, and the ground is scabbed with charcoal. Even the familiar sounds change: fewer birds, fewer insects, the sharp crack of a dead branch instead of the muffled give of a living one. From a distance, the hillside can look like a clean erasure. Up close, it is messier—a litter of partially burned cones, patches where moss has survived in damp creases, and, here and there, green that seems almost rude in its persistence.

The underground world meets catastrophe differently. Fire consumes what it can reach. Heat is a blunt instrument, but it does not travel far through soil, especially when the soil holds moisture. The result is a strange vertical geography: the surface can be sterilized while a few centimeters below, life continues. Roots may be killed, but deeper roots can survive. Hyphae may be severed near the top, while cords in cooler layers remain intact. Spores, built for hardship, can ride out conditions that would destroy an active mycelium.

Foresters and ecologists sometimes talk about a *seed bank*: dormant seeds lying in soil, waiting for light and space. There is also, less romantically named, a fungal equivalent—a bank of propagules, the fungal material capable of starting new symbioses. It includes spores, resistant structures such as sclerotia (compact survival bodies), and fragments of living mycelium that can regrow. Because fungi can be invisible even when abundant, it is easy to underestimate how much of a forest's future lies stored as microscopic potential.

Catastrophic disturbance is the moment when the bank matters most. In a stable forest, a tree can afford to be choosy. It can maintain multiple fungal partners and adjust its carbon payments gradually, in the slow bargaining rhythm of seasons. After a crown fire, the system does not have the luxury of subtlety. The environment becomes harsh and simplified: sun bakes the soil; nutrients pulse in strange forms; water runs off quickly because the sponge of litter is gone; erosive wind scours exposed ground. A seedling establishing in that setting is not looking for the best possible partner; it is looking for any compatible partner that can function immediately.

That urgency changes which fungi dominate. Many of the fungi that flourish under mature forests are late-successional specialists: well adapted to shaded soils and steady carbon inputs from established trees, sometimes poor competitors in the raw, fluctuating conditions of a burn scar. In the first years after disturbance, different fungi often take the lead—those that are good at arriving fast, tolerating extremes, and colonizing young roots. Ecologists sometimes call such species *ruderal*, borrowing a term from botany for plants that thrive in disturbed ground.

A concrete example comes from studies of ectomycorrhizal fungi after severe fire, where researchers have found that some lineages persist and recolonize with striking speed. Sydney Glassman and colleagues (2016) reported that high-intensity fire can reduce ectomycorrhizal inoculum density and richness, with rare taxa particularly hard hit, yet the spore bank can remain robust enough that post-fire seedlings become colonized largely by a subset of disturbance-tolerant fungi. Genera like *Rhizopogon* and *Wilcoxina* often appear as early colonizers in such contexts, turning up repeatedly in burned landscapes as if they had been waiting.

Waiting is not a metaphor. Spores can persist in soil for years, sometimes much longer, and they are not passive. They represent a strategy: invest in dormancy so that when a partner returns, the fungus can re-enter the economy quickly. In a mature forest, a spore is a lottery ticket with long odds, because established fungi already occupy most roots and defend their territory chemically and physically. After fire, the odds change. The competitive field is reset. The fungus that can germinate and colonize quickly suddenly has the advantage.

This fungal bank is not only a list of names. It is a set of mechanisms, each with its own ecological meaning.

Spores are the most obvious. They are tiny capsules containing the possibility of a new fungal body. Some travel on wind, some on water, many on animals. Small mammals that eat truffles (the underground fruiting bodies of some ectomycorrhizal fungi) disperse spores in their droppings; the fungi reward them with calories and scent. This is not a minor woodland curiosity. After a fire, animal movement can determine which fungal partners reach a recovering patch, especially if the patch is far from unburned refuges. A spore bank already in the soil is therefore a kind of insurance against isolation: it reduces dependence on immediate recolonization from outside.

Surviving mycelium is another mechanism, less dramatic but often crucial. If deeper soil stays cool enough, existing hyphal cords can remain alive, even if the surface network is badly damaged. When new roots appear, those living hyphae are already in place to colonize them, a shortcut that can save seedlings precious time. The image of a forest as something that must start over after fire is misleading. Often the forest is rebooting from a partially intact system.

Living host refugia provide yet another route. Some plants resprout after fire, retaining living roots and therefore maintaining symbiotic fungi. In arctic and boreal systems, resprouting shrubs can play this role. Rebecca Hewitt and colleagues (2013) showed that mycorrhizal fungal communities can be remarkably resilient after wildfire when resprouting hosts persist, acting as reservoirs that maintain compatibility across severity gradients. The lesson is not limited to the Arctic. Whenever some hosts survive, they act as bridges across disturbance: they keep fungi fed and present, and they hold the place biologically so that the return of trees is not a recolonization of empty ground but an expansion into a living neighborhood.

Even coarse woody debris—a fallen log, half-burned and damp inside—can function as a refuge. A nurse log is usually described as a moist seedbed, but it is also a microbial ark. In the protected interior, temperature fluctuations are muted and moisture persists. Seedlings establishing on nurse logs often encounter a different fungal community than seedlings on exposed mineral soil, and that difference can matter for survival.

If the fungal bank is so helpful, why do post-fire landscapes sometimes fail to recover? The bank is not a guarantee; it is a resource, and resources have limits.

Severity matters. A low-intensity fire that leaves the litter layer partly intact is not the same disturbance as a high-intensity fire that consumes organic horizons and heats soil deeply. When heat penetrates farther, it can kill not only surface hyphae but also propagules and fine roots. The fungal bank can be depleted, leaving seedlings to rely on long-distance dispersal. That dispersal is not always swift. If a burned patch is large and far from surviving forest edges, compatible spores may arrive slowly, and seedlings may spend critical months without proper symbiotic support.

Distance matters for another reason: dispersal is not uniform across species. Some fungi produce abundant wind-dispersed spores; others rely more on animals; others invest in fewer, larger propagules. A post-fire landscape does not simply need "fungi." It needs *the right fungi for the returning plants*. A pine seedling needs partners that can colonize pine roots. A spruce seedling needs a different compatibility set. The fungal bank is therefore part of what ecologists mean by *legacy*: the way past community composition shapes future community assembly.

There is also a tricky point about timing. Early colonizers are often not the same fungi that dominate mature forests. This is not a flaw. It is ecological succession belowground. Yet succession can stall. If early colonizers persist and block later arrivals, or if the plant community shifts toward species that maintain different fungal communities, the forest that regrows may not resemble the one that burned. Sometimes that is a healthy adaptation to new climate conditions. Sometimes it is an unwanted conversion, as when conifer forest gives way to shrubland because seedlings cannot establish at sufficient density. The fungal bank can influence which pathway wins simply by being more favorable to one set of plants than another.

These underground contingencies add emotional weight to the simple act of seeing a seedling. A seedling on a burn scar is not merely a brave plant. It is a biological negotiation taking place at the scale of micrometers. Its roots exude sugars and signaling molecules; fungi in the soil sense those exudates and respond; the first successful colonizers wrap the root tips and begin exchanging nutrients for carbon. Without that early handshake, the seedling's water relations can become unstable and its access to phosphorus can be too limited for rapid growth. In a post-fire world where summer droughts are sharper and soils are more exposed, speed matters.

This is why some restoration efforts have proposed adding mycorrhizal inoculum to burned sites. The idea has an intuitive appeal: if fungi are essential partners, why not supply them directly? Yet the fungal bank complicates this intervention in two ways.

First, inoculation may be unnecessary in some contexts. If the soil already contains abundant compatible spores and surviving mycelium, adding more may not change outcomes, and could even disrupt local communities by favoring introduced strains. Glassman's work (2016) has been interpreted as a cautionary note here: where spore banks remain and host seed sources are present, the system may already have the microbial ingredients for recovery.

Second, inoculation is not one thing. Commercial inocula vary widely, and the fungi that are easy to culture and mass-produce are not always the fungi most relevant to local forests. Introducing a non-local symbiont can create a short-term boost in greenhouse trials and still be ecologically inappropriate in the field. The more honest conclusion is that the fungal bank is a landscape feature to be assessed, not a blank to be filled reflexively. In some landscapes, especially where fires are severe and repeated, or where soil has been heavily disturbed after fire by erosion or salvage logging, the bank may indeed be depleted, and recovery may be limited by symbiotic availability. In others, the bank is quietly sufficient.

A historical anecdote helps frame how our thinking changed. For much of the nineteenth century and well into the twentieth, fungi were treated as accessories to plant life—sometimes beneficial, often pathological, usually ignored. Early forestry focused on trees as timber and soil as substrate. The recognition that mycorrhizas are widespread and often essential altered that worldview. By the late twentieth century, with isotope tracing and molecular methods, it became possible to treat the underground symbiosis as a measurable system rather than a vague concept. That shift is part of why discussions of post-fire recovery now include fungi at all. When a burned forest fails to regenerate, the explanation is no longer limited to "bad seed year" or "too dry." It can include "not enough compatible symbionts survived," a sentence that would have sounded strange to many foresters a century earlier.

The fungal bank also changes the way one interprets resilience. Resilience is often imagined as toughness: the ability to resist change. But in living systems, resilience is frequently about *capacity to reorganize*. A forest that returns after disturbance is not restoring every thread. It is building a functional network quickly enough to stabilize soil, cycle nutrients, and create shade. Early-successional fungi are part of that rapid assembly. They may be less efficient at long-term nutrient mining from mineral soil than late-successional partners, but they can be efficient at colonizing and supporting young roots under stress. Their value is measured not in perfection but in speed and adequacy.

One can see the bank's logic most clearly in a thought experiment. Imagine two burned hillsides. On the first, a mosaic of unburned patches remains: damp gullies, rocky outcrops, a few living shrubs that resprout. In the soil, spores are abundant and animal trails bring new ones. On the second, the burn was uniform and severe, organic horizons consumed, topsoil eroded by the first rains. Both hillsides receive the same seed rain. The first hillside becomes speckled with seedlings that survive their first summer, then their second. The second hillside has seedlings too, but many yellow and stall, or vanish after drought. The difference can be read aboveground as "microclimate" and "water." It can also be read belowground as "legacy and connectivity": the first hillside had an intact or quickly reassembling symbiotic infrastructure; the second did not.

That infrastructure is not a single network in the dramatic sense of one giant organism spanning a forest. It is a patchwork of survivorship and recolonization: surviving hyphae in deeper layers; spores waiting in cooler pockets; fungi maintained on resprouting hosts; new arrivals carried by wind and animals; the slow rebuilding of redundancy as more plant roots appear and more fungal partners establish. The forest's return is therefore not a simple reversal of destruction. It is an emergence from a bank of biological options stored in soil.

Standing on a burn scar, it is natural to look for the first green shoots and to treat them as the beginning. Often the beginning has been there all along, in the dark layer where heat did not reach, in a spore that endured by doing nothing, in a living root that resprouted quietly weeks after the flames. When the first seedling survives its first dry spell, it is drawing not only on its own reserves but on an underground inheritance that makes survival plausible: a partner that had already learned this soil, and had already learned how to wait.

Chapter 8

The Severed Link

Forests are not static collections of timber but dynamic conversations held together by fragile threads. When we strip the canopy or flood the soil with chemicals, we do not merely remove trees; we silence the underground dialogue that sustains the ecosystem's future. This chapter uncovers how modern interventions unravel millennia of fungal negotiation, transforming robust webs into fragmented ghosts.

8.1 When the Axe Falls

A forest looks like a collection of trunks, each tree neatly separated from the next by air. Forestry plans often inherit this view: the stand is a set of stems; the harvest is a subtraction problem. But underground, the forest is closer to a fabric than a filing cabinet. Roots overlap, touch, fuse, and compete. Mycorrhizal fungi—the symbiotic fungi that sheath or enter roots and trade minerals and water for plant sugars—extend that fabric into the soil. Their hyphae (microscopic filaments) are not merely accessories on the side of a tree; they are part of how a tree becomes a tree in the first place, particularly in nutrient-poor, acidic, or cold soils where direct root uptake is slow and costly.

Clear-cutting is a peculiar kind of violence to this fabric because it is not aimed at a single organism. It is aimed at the canopy as a whole, and therefore at the principal energy source that keeps the underground economy running. The stump, the slash, and the bare ground left behind can look like a reset, as if the forest has been returned to a neutral baseline. The fungi do not experience it as neutral. They experience it as famine, amputation, and sudden climate change all at once.

The famine is straightforward. Most mycorrhizal fungi live on carbon paid to them by living roots—not the carbon locked in wood, but the ongoing stream of sugars produced by photosynthesis. Remove the canopy and the stream stops. Some roots die quickly; others limp on briefly using stored carbohydrates, like a household spending down savings after losing a job. The fungi attached to those roots face an equally abrupt budget crisis. Hyphae in the soil are living tissue; they must be maintained, repaired, defended, and replaced. When carbon becomes scarce, the fungus retracts from the fringes. It abandons distant foraging lines. It stops investing in fine exploratory threads, and the network begins to thin, like a city turning off streetlights block by block during a power shortage.

The amputation is both literal and spatial. A mycorrhizal network is not one continuous organism draped under the forest; it is many organisms whose hyphae can interweave and whose spores and resting structures sit in wait. Still, continuity matters. Hyphae bridge soil pores, bind particles, link root tips, and traverse centimeters or meters to reach fresh resources. Heavy machinery compacts soil, collapses pores, and shears the fragile filaments. Ruts fill with water; raised tracks dry out. Fungal tissue is left hanging in microclimates it did not evolve to tolerate.

Then comes the climate change, but at the scale of centimeters. A shaded forest floor maintains a mild, buffered environment. Clearcuts turn the ground into an exposed surface that heats and cools rapidly, dries in wind, and receives rain as impact rather than mist.

Even if the soil a few inches down stays cool, the uppermost layers—where fine roots and mycorrhizal tips concentrate—become harshly variable. Fungi are sensitive to these fluctuations. A hyphal thread is a long, thin cell; its survival depends on a stable balance of water and solutes. When the topsoil alternates between baking and saturation, fungal communities shift toward the species that tolerate disturbance and away from those that built the old forest.

That shift is one of the subtler truths about clear-cutting, and it is easy to miss because it hides behind a deceptively reassuring number. If you dig up fine roots in a regenerating stand and look for mycorrhizas, you may still find plenty of colonized tips. The partnership does not vanish overnight. What changes, often dramatically, is *who* is doing the colonizing. The old network may be gone in the sense that the particular fungal species and strains adapted to the mature stand have declined, replaced by a different cast more suited to open ground, high light, and flushes of simple nutrients. The forest has not merely lost biomass; it has lost memory.

A useful way to picture this is to imagine a library. A clear-cut does not burn every book. Many shelves remain. But the catalog is scrambled, rare volumes disappear, and the librarians who knew where everything was have been dismissed. The building still looks like a library from the outside, yet the work it can do is different. Some fungi are rapid colonizers: they arrive early, tolerate heat and drought, reproduce quickly, and exploit simple, abundant resources. Others are slow specialists: they knit long-lived, far-reaching hyphal systems, access complex organic nutrients, and form stable associations with particular tree lineages. A mature forest needs both kinds, but it is the slow specialists that tend to anchor continuity across decades. Clear-cutting tilts the balance toward the fast and the general.

The human history of clear-cutting makes the biological history easier to understand. In the nineteenth century, as industrial logging expanded in North America and parts of Europe, clear-cutting fit the machinery and the markets. Sawmills wanted uniform logs delivered quickly. Railways and rivers carried large volumes efficiently. The forest was treated like a stockpile: cut, move on, repeat. In many regions the result was not simply a second-growth forest but an altered landscape that burned more readily, eroded more quickly, and regenerated unevenly. People noticed the bare hillsides and the floods long before they had words for mycelial continuity. A famous early warning came from French engineers and foresters who linked mountain deforestation to destructive floods and landslides; reforestation programs in the Alps and Pyrenees were as much about stabilizing watersheds as about timber. They were responding to a surface symptom of an underground loss: when trees and fungi no longer bind soil and regulate water, the land becomes louder—storms translate more directly into disaster.

Modern forestry, with its science, regulations, and replanting, is not the nineteenth century. Yet the old logic still echoes: efficiency encourages simplification. A clear-cut is easy to plan, easy to harvest, and easy to replant. And because tree seedlings can be made to grow, it can appear that the system is resilient. But the seedlings that survive may do so in a soil that has been biologically rearranged. Their growth, their drought tolerance, and their resistance to disease can depend on the invisible help that the old forest provided and the clear-cut disrupted.

Consider a concrete scene from the Pacific Northwest, where conifers such as Douglas-fir have been managed intensively for more than a century. A freshly logged unit is replanted with rows of seedlings. From above, it looks like renewal. From below, the seedlings enter a soil where the most extensive ectomycorrhizal networks (the type of mycorrhiza common in many temperate and boreal trees, forming a sheath around roots) have been cut off from their main carbon donors. The fungal community that establishes on the seedlings may come from spores blown in, from resistant structures already in the soil, or from surviving patches at the edges. If the clear-cut is large and the edges are far away, the colonists are more likely to be those that disperse well and tolerate exposure—not necessarily those that best support long-term forest function. What the seedlings inherit is not the mature forest's underground infrastructure but a kind of provisional wiring.

This is where the idea of *legacy trees* enters, and with it a quiet shift in how logging can be imagined. A legacy tree is a living remnant intentionally retained through harvest: a large old tree left standing, or a small group left as a patch, sometimes along streams or ridgelines, sometimes scattered like islands. Foresters may leave them for wildlife habitat, aesthetics, seed sources, or cultural reasons. From the fungal perspective, their most important role can be simpler and more fundamental: they keep paying the underground bills.

A living, photosynthesizing tree does not merely occupy space. It keeps carbon flowing into roots. It keeps root tips active. It maintains the mycorrhizal fungi that depend on those roots, and it maintains a humid, shaded microclimate beneath its crown. In a clear-cut, a legacy tree becomes a kind of refugium—a refuge from the harshness of open ground. The fungi associated with it are not guaranteed to be identical to those that dominated before harvest, and the benefits spread only so far. Still, the legacy tree functions like a lifeboat: not a full restoration of the lost ship, but a place where living networks persist through the disturbance and can expand outward again as the forest regrows.

The lifeboat metaphor is not sentimental. It is a description of survival strategy. Many mycorrhizal fungi can persist as spores or dormant structures, but these are often a gamble. A spore is a message in a bottle: it may land in the right place, at the right time, with the right host nearby, and then succeed against competitors. By contrast, an actively fed mycelium—a living network attached to living roots—is an operating business. It has employees (hyphae), equipment (enzymes), and supply lines (carbon). When conditions improve—when seedlings' roots approach, when shade returns, when litter accumulates—it can respond quickly. It can colonize new roots not from scratch, but from an established front.

How far can that influence extend? Not indefinitely. Hyphae are not highways spanning clear-cuts like suspension bridges. Their reach depends on species, soil moisture, and the availability of roots to connect. In practice, the lifeboat effect is likely strongest within and near the retained patch, fading with distance into the open unit. That spatial limitation is exactly what makes the design of retention forestry matter. A single legacy tree in a vast clear-cut is better than none, but it is also an island surrounded by biological surf. A cluster of retained trees creates a larger, more stable interior microclimate and more root biomass to sustain fungi through the vulnerable years.

A crucial detail is that not all retention is equal from the fungal point of view. Leaving dead wood and stumps has value, particularly for saprotrophic fungi (those that decompose wood) and for moisture retention. But the mycorrhizal lifeboat depends on *living* roots. A stump is a memorial; it is not a sugar source. In the months after harvest, some stumps and roots remain physiologically active for a time, especially if connected to neighboring living trees through root grafts or shared mycorrhizal partners. But as the remaining photosynthetic tissue is removed, the capacity to feed fungi dwindles. The lifeboat that matters most is the one that still eats.

There is also a human temptation to over-romanticize what that lifeboat can do. Popular retellings of forest networks sometimes suggest a benevolent underground internet in which old trees lavish resources on seedlings. The reality is more constrained and more interesting. Carbon can move through shared fungal connections under field conditions, and shading or other stress can influence the direction and amount of movement. Yet a clear-cut is not simply a moral drama of generosity and abandonment. It is a physiological system reacting to gradients of light, water, and nutrients. When a legacy tree helps a seedling, it may do so indirectly: by maintaining fungi that improve nutrient uptake, by moderating microclimate, by stabilizing soil structure, or by keeping pathogens in check through competitive fungal communities. These are not cinematic acts, but they are powerful ones.

The idea that legacy trees serve as biological reservoirs has practical echoes in Indigenous and local forest knowledge. In parts of the Pacific Northwest, for example, communities valued certain large, long-lived trees for reasons beyond timber: cedar for cultural materials, old trees for landmarks, groves for habitat. Even when the language was not about mycorrhizas, the practice of retaining important trees and patches preserved continuity. A forest managed as a place to return to, rather than a place to empty, tends to keep more of its living infrastructure intact. The fungal web is one beneficiary among many.

The most dramatic losses after clear-cutting can be the losses you cannot replant. Seedlings can be replanted. Soil can be scarified or amended. Roads can be decommissioned. But certain fungal communities take time to reassemble, and some may not reassemble at all if the site is repeatedly disturbed or converted to simplified rotations. The legacy of harvest can persist for decades, not necessarily as a barren underground, but as a different underground. This matters because fungi are not interchangeable tools. Different mycorrhizal fungi explore soil differently: some form dense mats; some send out long-distance cords; some specialize in mining nitrogen locked in complex organic matter; others are more dependent on readily available forms. The composition of the fungal community influences how nutrients cycle, how seedlings establish, and how forests respond to drought and disease.

Retention forestry is sometimes described as a compromise between clear-cutting and selective logging, but biologically it can be more than a compromise. It is a way of admitting that continuity itself is a resource. Keeping patches of the old forest alive keeps the web alive, and a living web can knit back into disturbed ground faster than a web that must be reinvented spore by spore.

A final complication makes the story less tidy and more realistic. Clear-cutting does not simply remove trees; it also changes the competitive landscape among fungi. When mycorrhizal fungi decline, saprotrophic fungi and bacteria may increase in relative importance, feeding on the pulse of dead roots and logging residues. Some of these decomposers are essential for recycling nutrients and rebuilding soil organic matter; others may accelerate carbon loss under the newly exposed conditions. The underground is not a single network with a single purpose. It is a crowded marketplace of organisms responding to sudden opportunity and sudden scarcity. The question is not whether life continues—it does—but whether the particular forms of life that make forests durable can persist through the disturbance.

The lifeboat image helps here because it suggests a kind of triage. No one expects a lifeboat to carry everything. What it carries are the living lineages that can rebuild. Legacy trees and retention patches are not a guarantee of ecological innocence. They do not erase the shock of harvest. But they preserve connection: the ongoing exchange between canopy and soil, the humid shade that shelters fine roots and fungal tips, and the local presence of fungal partners already adapted to that place.

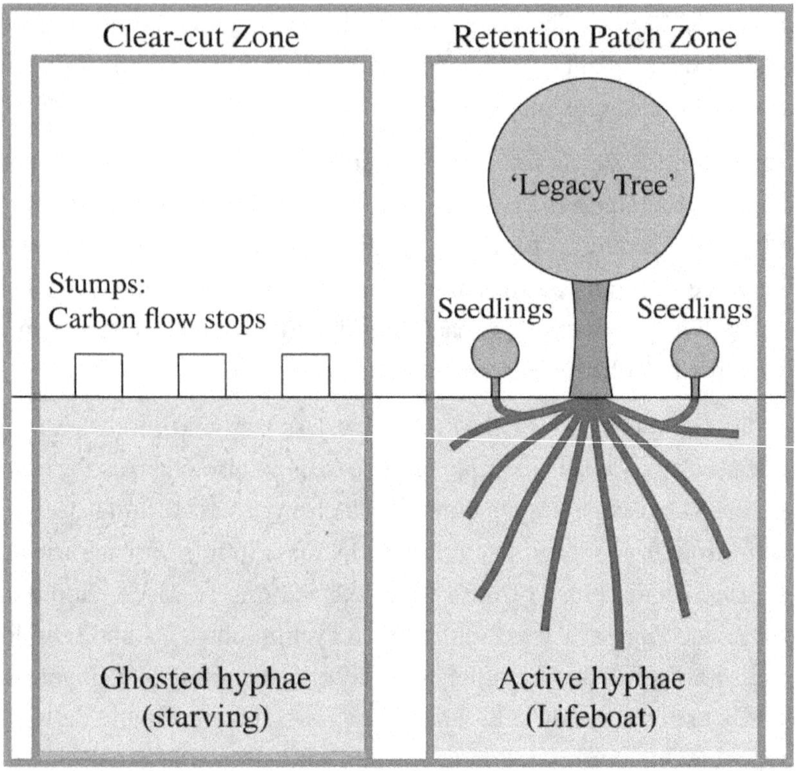

A clear-cut, then, is not only the removal of wood. It is the sudden interruption of a living transaction: carbon for nutrients, sugar for water, photosynthesis for soil mining. Leave no living trees and the interruption becomes near-total, forcing the next forest to assemble its underground alliances from fragments, spores, and chance arrivals. Leave a few well-placed elders standing and something continuous survives: a fed mycelium, a humid pocket of soil, a local archive of partners already fluent in the chemistry of that place. When the first roots of the regenerating forest reach those pockets, they do not meet bare ground. They meet a web that, though strained and shortened, is still alive enough to take their hand.

8.2 The Nitrogen Paradox

Nitrogen is the substance most likely to make a forester, a farmer, and a factory manager nod in agreement. It is the element of protein and chlorophyll, the bottleneck that keeps plants small and pale when soils are poor. Add nitrogen, and leaves darken; growth quickens. For more than a century, that basic response has been turned into policy and industry: fertilize to increase yields, engineer nitrogen into forms that plants can take up, and accept the green surge as proof that the ecosystem is healthier.

The paradox begins when extra nitrogen arrives without asking, and in quantities the forest did not evolve to negotiate. It comes down as invisible fertilizer in rain and dust. It drifts in from highways, power plants, and livestock operations. It is applied deliberately to plantations to push timber volume. The canopy may indeed look greener. Yet beneath the litter, the relationship that allowed trees to thrive on lean soils can start to loosen, as if a handshake has been replaced by a brief, distracted tap of fingers.

The relationship in question is the mycorrhizal partnership: the trade between plant and fungus. The plant provides carbon—sugars produced by photosynthesis—and in return the fungus provides services the plant cannot easily perform alone, especially in soil. Those services include scavenging nitrogen and phosphorus, transporting water, and exploring pore spaces too small for roots. In ectomycorrhizal associations (common in pines, oaks, beeches, birches, and many other temperate and boreal trees), the fungus forms a sheath around root tips and sends hyphae outward into the soil, often specializing in pulling nitrogen out of complex organic matter. In arbuscular mycorrhizal associations (common in many maples, ashes, and tropical trees, and in most herbaceous plants), the fungus penetrates root cells and forms tiny exchange structures, often excelling at phosphorus uptake. These are not hard rules, but they are useful tendencies. A forest can be dominated by one style or contain a mosaic of both, and the balance matters for how nutrients cycle and how carbon is stored.

A tree does not keep a fungal partner out of kindness. It pays for a service. The first step in understanding nitrogen's disruptive power is to take that trade seriously as economics, in the plain sense of budget and incentive. The tree's carbon is not free; it is the tree's potential height, seeds, defense chemicals, and stored reserves. If nitrogen becomes abundant in the soil in forms the tree can access without fungal help, the price of the fungal service rises relative to its benefit. The tree can reduce payment. It can allocate less carbon to roots, produce fewer fine roots, or simply starve certain fungal partners. The fungus, suddenly unpaid, either retreats, switches strategies, or is replaced by a competitor that can survive on less.

That sounds almost too rational, as if a tree is doing arithmetic. In practice it is physiology. When nitrogen is plentiful, plants often reduce the carbon they send belowground. Hormones and nutrient signals alter root growth and the quantity and composition of sugars exuded into the soil. Mycorrhizal colonization can decline, or it can remain but with a different cast of fungal species: those that ask for less carbon, those that reproduce quickly, those that behave more like opportunists than long-term contractors. The old network does not necessarily vanish; it becomes a thinner, less diverse version of itself, and in some places it is replaced by a different kind of fungal community altogether.

This helps explain why nitrogen pollution can harm forests even as it makes them look vigorous. When the fungal web is simplified, the forest can lose capacities that are not obvious in a single growing season. Diverse ectomycorrhizal communities contain species with different foraging habits. Some send out short hyphae that comb through nearby soil. Others build thicker, longer-distance cords that function like pipelines, moving resources across patches. Some are better at accessing nitrogen bound in proteins and other complex molecules; others specialize in more readily available forms. Under chronic nitrogen deposition, certain ectomycorrhizal types are often the first to disappear: those that invest in extensive, hydrophobic (water-repelling) hyphal systems and those that deploy enzyme toolkits for mining organic nitrogen. These are expensive strategies, paid for by plant carbon. If the plant stops paying, the strategy collapses.

The result can feel backwards: more nitrogen in the environment, but less capacity to use nitrogen that is locked away in the forest's own organic matter. A forest is not a hydroponic system where nutrients arrive in perfectly soluble form. Much of the nitrogen is tied up in dead leaves, wood fragments, and the bodies of microbes. Ectomycorrhizal fungi are among the organisms that help unlock this store, either directly through enzymes or indirectly by shaping the microbial community around roots. If chronic deposition shifts the ectomycorrhizal community toward species that do less of this work, the forest becomes more dependent on the ongoing drip of external nitrogen. It is a subtle kind of dependency, like a city that stops maintaining its reservoirs because imported water seems cheap, only to discover later that imports are unreliable and the local system has decayed.

A concrete example comes from the long arc of industrial air pollution in Europe and eastern North America. In the mid-twentieth century, coal combustion and industrial processes poured nitrogen oxides and sulfur compounds into the air. The sulfur story became notorious as "acid rain," and regulations in the late twentieth century reduced sulfur emissions dramatically in many countries. Nitrogen proved more stubborn, in part because vehicles and intensive agriculture continued to emit it. Forest scientists watching these regions noticed that nitrogen deposition did not just fertilize trees; it altered soil chemistry, leached base cations, and pushed some ecosystems toward "nitrogen saturation," where the soil and vegetation could no longer retain all the incoming nitrogen. Streams carried nitrate out of watersheds. The forest, in effect, was being force-fed beyond its capacity to metabolize. In such conditions, changes in mycorrhizal communities were not a side note; they were part of the mechanism by which the forest's internal regulation failed.

One way to feel the paradox emotionally is to imagine a child given unlimited sweets. There is energy, excitement, and rapid change. Over time, teeth rot and appetite for proper food diminishes. The damage is not always obvious until something breaks. With nitrogen, the "sweets" are reactive forms like nitrate and ammonium delivered steadily from the sky or spread deliberately. The forest responds with growth pulses, but the mutualisms that evolved to cope with scarcity become less necessary, and therefore less supported.

Forests differ in how quickly they reach this point. Some are genuinely nitrogen-limited and can absorb extra inputs for years with modest harm, especially if other nutrients (like phosphorus) are also available. Others are limited by phosphorus or by the ability of roots and fungi to access water, and in those forests nitrogen additions can behave like an irritant: they push growth in a way that increases demand for other resources, tightening constraints rather than loosening them. A tree encouraged to build more leaf area by nitrogen must then supply that leaf area with water in summer and with phosphorus for membranes and energy transfer. If the mycorrhizal community that helps with those constraints has been weakened by the same nitrogen influx, the tree can become paradoxically more vulnerable.

This vulnerability shows up most clearly under stress. A mature forest can appear stable until drought arrives, or until a pathogen takes advantage of weakened defenses. Mycorrhizal fungi are not a shield against all enemies, but they contribute to resilience in ways that can be hard to measure in the moment: improved water uptake, access to nutrient patches, and the maintenance of a diverse microbial neighborhood that competes with root pathogens. When chronic nitrogen inputs simplify that neighborhood, the forest's response to stress can become more brittle.

The paradox also plays out at the scale of landscapes. Nitrogen deposition does not fall evenly; it forms gradients downwind of cities, highways, and agricultural regions. Those gradients can reshape the very composition of forests over decades, favoring trees that partner primarily with arbuscular mycorrhizal fungi over trees that depend heavily on ectomycorrhizal partners. The reasons are still being worked out in detail, but the pattern has been observed at broad scales: chronic nitrogen inputs can coincide with a shift away from ectomycorrhizal-dominated systems. If that shift occurs, the underground economy changes character. Ectomycorrhizal fungi and arbuscular mycorrhizal fungi differ in how they acquire nutrients and in how they influence decomposition and soil carbon storage. A landscape that flips its dominant mycorrhizal mode is not just swapping one set of fungal species for another; it is changing the rules of nutrient cycling and, potentially, how much carbon is retained in soils.

That last point matters because nitrogen is often discussed as if it were automatically climate-friendly: more nitrogen, more plant growth, more carbon captured. But a forest is not only a pile of wood. A large fraction of its carbon lives underground as roots, fungal biomass, microbial residues, and stabilized organic matter. Mycorrhizal fungi influence how carbon moves into and through those pools. In some contexts, nitrogen additions can reduce microbial biomass and slow decomposition, which might increase soil carbon. In other contexts, nitrogen can suppress certain fungal activities while favoring others, altering the balance between carbon inputs and outputs. Experiments comparing mycorrhizal "strategies" suggest that arbuscular-mycorrhizal and ectomycorrhizal systems can differ in how they affect soil carbon losses, in part because they couple nitrogen uptake to different microbial responses. The overall outcome is not a simple arrow from nitrogen to carbon storage; it depends on which partnerships dominate and how those partnerships shift.

A dinner-party way to put it is this: forests do not simply need nitrogen. They need a functioning negotiation between living roots and soil life. Flood the system with one resource and the negotiation changes. The forest may become temporarily richer in the currency of nitrogen, while becoming poorer in the social infrastructure that made that currency usable across seasons and stresses.

There is another, more immediate way nitrogen severs the bond, and it is chemical rather than economic. Reactive nitrogen can acidify soils when ammonium is converted to nitrate, releasing hydrogen ions. Acidification can mobilize aluminum in forms that are toxic to roots, and it can leach calcium and magnesium—nutrients trees need for cell walls and for the machinery of photosynthesis. Mycorrhizal fungi can buffer some chemical stresses, but they are also vulnerable to them. A soil that becomes more acidic and more chemically imbalanced can select against sensitive fungi, compounding the effect of reduced carbon payments. The tree, in turn, may struggle with nutrient imbalances that nitrogen alone cannot fix. The gift becomes a trap: a growth signal without the mineral foundation to support that growth.

Fertilization in managed forests brings the paradox into sharp relief because it is intentional. A plantation manager may apply nitrogen to increase timber volume, sometimes with impressive short-term gains. From the standpoint of annual growth, the intervention works. From the standpoint of fungal partnership, the intervention can be a shock, especially if repeated. The fungal community can shift toward a smaller set of nitrophilic species—species that thrive under high nitrogen—while species that specialize in scavenging scarce nutrients decline. If fertilization is heavy and sustained, microbial biomass in the soil can decrease, respiration can change, and the ratio of fungal to bacterial activity can shift toward bacteria. The soil becomes less like a fungal-dominated, slow-cycling system and more like a bacterial-dominated, fast-cycling system. That change is not inherently "bad" in all contexts, but it is a real transformation of the underground world, and it is not what most people picture when they hear the word fertilization.

A historical anecdote makes this transformation tangible. In the years after the Second World War, the same industrial capacity that had produced explosives and other wartime chemicals was redirected toward fertilizer production. The Haber–Bosch process, which fixes atmospheric nitrogen into ammonia using high pressure and energy, became a foundation of modern agriculture and, indirectly, of modern landscapes. It is hard to overstate what this meant: humans learned to manufacture biologically available nitrogen at enormous scale. Crop yields soared; famine risks shifted. Less often remembered is that the nitrogen did not stay politely on fields. It volatilized, ran off, and drifted. Forests downwind received a slow-motion experiment in enrichment. In some regions the first sign of trouble was not in the canopy but in streams, where nitrate levels rose. Only later did the underground story become clearer: the microbial and mycorrhizal communities that had evolved under tight nitrogen budgets were being pushed into a different regime.

If nitrogen is the paradoxical gift, "critical loads" are the attempt to put a number on when the gift becomes harmful. A critical load is, roughly, an amount of pollutant deposition below which significant harmful effects are not expected for sensitive elements of the ecosystem. The exact values vary by forest type, soil, climate, and history. Conifer forests with ectomycorrhizal communities adapted to nutrient-poor conditions are often among the most sensitive. The usefulness of the concept is not in giving a universal threshold but in forcing a practical question: how much enrichment can a forest absorb before it begins to lose the fungal partners that help it function?

The unsettling answer is that the damage can begin long before the trees look sick. Mycorrhizal communities can change quietly. Certain fungi vanish from root tips. The fine structure of the hyphal web becomes less extensive. The enzymatic capabilities in the soil shift. In the language of human health, the forest's bloodwork changes before symptoms appear.

None of this means that nitrogen is a villain in itself. In a genuinely nitrogen-poor stand, modest additions can improve growth and even support recovery after disturbance. Nor does it mean that mycorrhizal fungi always suffer under added nitrogen; some will flourish. The trouble lies in chronic, high inputs that rewire incentives and chemistry for years, nudging forests away from long-coevolved partnerships toward a simplified set of alliances that are optimized for abundance rather than for balance.

The paradox becomes most vivid when one imagines the forest as a set of contracts written in carbon. A tree signs those contracts every day by allocating sugars belowground. The fungi fulfill them by extending the tree's reach into soil. When nitrogen rains down freely, the tree rewrites the contracts, paying less for the same service or choosing cheaper contractors. The forest does not notice immediately because the canopy can still look luxuriant. But when drought tightens, when pathogens probe, when other nutrients become limiting, the missing clauses matter. A forest that has reduced its investment in fungal networks has also reduced its options.

In places where nitrogen deposition has begun to fall due to cleaner air policies, the question becomes whether the fungal web can recover and, if so, on what timescale. Recovery is not simply a matter of turning off the tap. It depends on whether sensitive fungi still have refuges, whether soils have been acidified and depleted of base cations, and whether the tree community itself has shifted toward species that do not support the same fungal partners. The forest remembers enrichment as surely as it remembers fire or logging: not as a single event, but as a changed baseline.

The most honest way to hold the paradox is to keep both truths in view at once. Nitrogen can make trees grow, and nitrogen can make forests less secure. The first truth is easy to celebrate because it is visible. The second is harder because it belongs to the world beneath our feet, where loss often looks like mere rearrangement. Yet a rearrangement of the fungal community is not cosmetic. It is a change in who negotiates nutrients, who pays whom, and how much of the forest's strength is stored in relationships rather than in wood.

8.3 Unlocking the Carbon Vault

Climate conversations like to stand upright. They point to trunks, canopies, and the visible drama of forests: the standing carbon in wood, the loss when trees are felled or burned, the regrowth that pulls carbon dioxide back from the air. The underground story is less photogenic and, for that reason, easier to neglect. Yet the great majority of a forest's carbon is not arranged as a neat inventory of logs. It is dispersed through soil as crumbs, coatings, and ghostly residues: molecules bound to minerals, fragments locked inside aggregates, the remains of roots and fungi transformed by microbes into forms that can persist for decades or centuries. This is the carbon vault, and it is not a passive storehouse. It is a structure built and maintained by living processes.

Mycorrhizal fungi sit at the hinge of this vault. They are conduits for carbon flowing from leaves to soil, and they are engineers of soil structure. They also influence which decomposers thrive, which enzymes are produced, and how quickly organic matter is processed. Disrupt the fungal network, and the vault does not merely lose a bit of capacity. Its doors can be left ajar.

The temptation is to picture soil carbon as leaf litter that simply fails to rot. That image is comforting because it suggests a straightforward solution: stop disturbing the soil, and carbon will sit safely, like a book on a shelf. Modern soil science offers a more unsettling, but more useful, picture. Much stable soil carbon is not preserved because it is inherently tough; it is preserved because it has been transformed and tucked away. Microbes take in plant carbon, build their bodies, and when they die their remains become microbial necromass—a kind of molecular compost that binds to minerals and becomes difficult for other organisms to access. The vault is filled by metabolism as much as by accumulation. That means it can be emptied by disrupting metabolism, too, especially the metabolism that depends on steady carbon payments from living plants.

Roots are central here, and so are the fungi attached to them. In some boreal systems, careful radiocarbon work exploiting the "bomb" pulse of ^{14}C from mid-twentieth-century nuclear testing has suggested that a surprisingly large share of long-term stored soil carbon originates from roots and root-associated microorganisms rather than from aboveground litter alone. This is not mystical: roots leak carbon, roots die, fungi receive carbon and build biomass, and microbial communities process all of it into stabilized forms. The visible fall of leaves is only part of the input stream.

Once you start thinking this way, the earlier disturbances take on an additional meaning. When the axe falls and the canopy is removed, the most immediate change underground is not simply fewer roots next year. It is a sudden reduction in the daily infusion of carbon that kept root tips active and mycorrhizal fungi fed. A mycorrhizal network can be extensive beyond what the eye can imagine: extramatrical mycelium, the hyphae extending outward from roots into soil, can form a vast, thin biomass spread across meters. It turns over: threads grow, senesce, are grazed by microarthropods, and are decomposed. That turnover is not waste; it is one of the ways carbon becomes incorporated into soil in small, stabilizable pieces. Cut off the carbon supply and the fungal mycelium retracts and dies back. A pulse of dead fungal and root tissue enters the decomposer system at the same time the soil becomes warmer, drier, and more oxygenated near the surface. The vault is suddenly being fed a kind of kindling under conditions that encourage burning.

The "burning" is microbial respiration: carbon converted back to CO_2. It can accelerate for reasons that are both physical and biological. Physical disturbance breaks soil aggregates—the clumps of particles that create protective microhabitats where organic matter can be shielded from enzymes and oxygen. Machinery, compaction, and drying-rewetting cycles can crack these clumps open, exposing formerly protected carbon to microbial attack. Biologically, shifting fungal communities can change the balance between organisms that build stable structures and organisms that mine carbon quickly.

It helps to picture soil aggregates as miniature safes. Each has a lock (the mineral surfaces that bind organic molecules), a door (the pore structure that limits who can enter), and guards (the microbial community that competes and regulates enzyme production). Fungal hyphae contribute to the door and the guards. They physically enmesh particles, and they produce sticky substances that help glue soil together. In arbuscular mycorrhizal systems, one of the famous candidates for such glue has been "glomalin," a term associated with arbuscular mycorrhizal fungi. The details of what glomalin is chemically have been debated, but the larger point remains robust: fungal growth promotes aggregation and the formation of stable microhabitats. Ectomycorrhizal hyphae, too, form mats and cords that bind soil and influence how water moves through it. When these structures are reduced, soil can become more erodible, more prone to crusting at the surface, and more susceptible to rapid decomposition pulses after wetting.

A historical anecdote makes this physical aspect easier to grasp. In the 1930s, when drought and ploughing combined to create the Dust Bowl in the United States, the image that traveled the world was airborne topsoil. The crisis is remembered as a failure of farming practices and a catastrophe of climate. It was both, but at its heart it was also a failure of soil structure. The "crumb" architecture of prairie soils, built over centuries by roots, fungi, and soil organisms, had been broken down. Once aggregation fails, soil is not merely dirt; it is a powder. Forest soils are different from prairie soils, yet the lesson transfers: structure is not guaranteed. It is earned. Disrupt the organisms that build it, and a landscape can lose carbon and fertility in a way that is hard to reverse quickly.

The fungal role in carbon is not only architectural. It is also strategic. Ectomycorrhizal fungi can affect decomposition in contradictory ways that depend on context. In some surface litter layers, ectomycorrhizal fungi appear to slow decomposition by competing with saprotrophs—the free-living decomposer fungi—for nitrogen and other nutrients. This phenomenon is often discussed as the Gadgil effect, after early work suggesting ectomycorrhizal fungi could suppress litter decay. If decomposition is slowed, more carbon might remain in litter, at least temporarily. Yet ectomycorrhizal fungi are not always decomposition brakes. Many produce oxidative enzymes and can interact with complex organic matter, especially in deeper organic layers. They can, under some conditions, stimulate the breakdown of certain compounds, or at least reshape where and how decomposition occurs. The effect is not a simple on/off switch; it varies with depth, moisture, nutrient availability, and the particular fungal species present.

This is why network disruption matters so much. A diverse ectomycorrhizal community contains a range of functional types: some better at building extensive hyphal networks, some better at enzyme production, some better at tolerating drought, some better at working in organic horizons. Disturbance and chronic nitrogen inputs can selectively remove certain types, simplifying the system. When the system is simplified, the balance between suppression and stimulation of decomposition can change, and so can the routes by which carbon becomes stabilized as microbial necromass.

Nitrogen enters this vault story in a way that initially feels counterintuitive. Nitrogen pollution can sever the mutualism by changing the tree's incentives, yet nitrogen additions can sometimes increase soil carbon stocks in long-term experiments. Both can be true. If added nitrogen suppresses microbial enzyme production or reduces decomposition of certain litter components, carbon can accumulate. If nitrogen increases plant growth and root inputs, more carbon enters the soil. If nitrogen fosters microbial biomass that later becomes necromass bound to minerals, the vault can, in a sense, be filled. Yet the same nitrogen regime may also simplify mycorrhizal communities, reduce the extent of extramatrical mycelium, and weaken soil structure over time, making carbon more vulnerable to physical disturbance or climatic extremes. A vault can be filled with more valuables even as its walls are weakened. The short-term balance sheet can look good while the long-term security deteriorates.

A concrete, contemporary example is the fate of peatlands and boreal soils under warming and disturbance. Peat stores enormous quantities of carbon accumulated under waterlogged conditions where decomposition is slow. In boreal forests, thick organic horizons can behave like a semi-peat vault, storing carbon in cold, acidic conditions. When such systems are drained, compacted, or warmed, decomposition accelerates and carbon losses can be large and persistent. Mycorrhizal networks are not the only actors here, but they influence how plants allocate carbon belowground, how roots explore, and how microbial communities are structured. If the mycorrhizal web is disrupted by logging, road-building, or chronic nitrogen, the system may become less able to rebuild stable organic horizons after disturbance, increasing the risk that warming turns the vault into a source.

There is a more intimate example you can perform with your hands. Pick up a handful of forest soil from an undisturbed patch and gently crumble it. Often it breaks into small aggregates that hold together, with fine roots and fungal threads visible as pale filaments. Now pick up soil from a heavily compacted track or an eroded, exposed spot. It may smear, or it may crumble into dust, lacking that fibrous cohesion. The difference is partly moisture, partly texture, but it is also biology. Fungal threads are literal threads. They bind. They connect. They create the micro-architecture that makes carbon difficult to steal back.

Disturbance unlocks the vault by altering three linked flows: the flow of carbon into the soil, the flow of oxygen and water through soil pores, and the flow of energy within microbial communities.

First, carbon inputs. Living canopies allocate carbon belowground continuously. Some becomes root biomass, some becomes exudates (sugary compounds secreted by roots), and some becomes mycorrhizal biomass. Remove the canopy or reduce photosynthesis through stress and the pipeline narrows. The living fungal web shrinks. The system becomes less able to invest in the slow construction of stable carbon pools.

Second, oxygen and water. Soil carbon is protected in part by limited access: microbes and enzymes need water films to move, and oxygen availability controls which metabolic pathways dominate. Disturbance changes pore structure. Compaction can reduce oxygen deep down, yet it can also increase runoff and drying near the surface. Clear-cuts can warm the soil and increase evaporation. Drying followed by rewetting can create pulses of microbial activity that release CO_2 in bursts, like a bellows. Aggregates that once held carbon in low-oxygen micro-sites can crack, making oxygen access easier. The vault's internal climate shifts.

Third, microbial energy balance. Microbes do not decompose everything equally. They allocate effort based on what is available and what nutrients constrain them. Mycorrhizal fungi can act as gatekeepers by competing for nutrients and by shaping which decomposers can thrive near roots. When the mycorrhizal community is weakened, decomposer communities may shift. In some cases bacteria become relatively more prominent compared to fungi, changing the types of residues produced and the pathways by which carbon becomes stabilized or lost. The vault is not only a physical structure; it is a negotiated truce among organisms with different appetites.

Because the carbon vault is built from innumerable tiny transactions, it is sensitive to repeated, modest disruptions. A single harvest or a single drought does not necessarily empty it, just as a single withdrawal does not bankrupt a bank. The danger lies in compounding: repeated clear-cuts on short rotations, chronic nitrogen deposition that alters partnerships, soil compaction that persists, and a warming climate that increases microbial metabolism. Each stressor nudges the system toward faster cycling and weaker stabilization.

It is worth pausing on the emotional weight of this, because the underground story can sound abstract. Carbon in soil is not an aesthetic pleasure in the way a cathedral-like canopy is. Yet it is one of the few forms of climate mitigation that is both large and local. Soil carbon influences water retention, buffering streams in drought and moderating floods. It influences fertility and therefore the ability of forests to regrow after disturbance. It influences the resilience of rural economies that depend on timber, recreation, and clean water. Unlocking the vault is not just a global accounting error; it is a loss of local stability.

The most unsettling feature of a carbon vault is that it cannot be rebuilt quickly once compromised. You can plant trees in a day. You cannot rebuild a century of microbial residues bound to minerals with a single intervention. The rebuilding requires time, steady inputs, and a community of organisms able to transform carbon into stable forms. Mycorrhizal fungi are among the organisms that make that transformation possible because they mediate the flow of carbon into soil and help create the physical and biological conditions under which carbon becomes protected.

This changes the moral of forest management in a quiet way. If forests are treated only as aboveground biomass, then disturbances are judged mainly by how quickly trees regrow. If forests are treated as coupled canopy—soil systems, then disturbances are judged by whether they maintain the living processes that build long-term stability. The fungal web becomes a practical concern, not a poetic one.

A tree ring records a year. Soil often records a century. The fungal network is one of the few living structures that touches both clocks at once: it responds within days to a change in carbon supply, and it shapes the slow formation of stable carbon pools over decades. When that network is severed, the forest loses not only a partner for nutrient uptake but an architect of its underground savings account. The vault does not explode in a single moment; it leaks. And in a warming world, a leak that persists is enough to change the balance from storage to loss, turning the ground beneath a forest from a quiet reserve into an unremarked source of CO_2.

Chapter 9

Healing the Soil

Restoration has long focused on what we see above ground, often ignoring the severed lifelines beneath that determine a forest's fate. By pivoting from industrial monocultures to network-aware forestry, we can harness the resilience of the fungal web to heal degraded landscapes. This shift requires not just planting trees, but transplanting the ancient alliances that sustain them.

9.1 The Architecture of Retention

A clear-cut looks like an ending. The trunks are gone, the canopy is gone, the birdsong thins out, and the sunlight falls too hard on the ground. What is less obvious is that a clear-cut is also a kind of amputation. The forest is not only the standing timber; it is also the living weave beneath it, where roots braid through soil and fungi braid through roots. When the trees are removed at once, that weave is torn in many places at once, and the damage does not stop at the stump line. Seedlings planted afterward may arrive with green needles and straight intentions, yet find themselves in a place where the essential relationships that make a forest function have been erased.

Retention forestry begins with a different instinct: not to treat a harvested stand as an empty stage that can be reset, but as a community whose survivors can keep the backstage running. The phrase *variable retention* can sound bureaucratic, as if it were a spreadsheet decision about how many stems to leave. In practice it is closer to an architectural principle. Leave enough living structure—and leave it in the right configuration—so that regeneration does not start from zero.

To see why the configuration matters, it helps to imagine what a seedling needs in its first years. It needs light, of course, and water, and the ability to keep its needles from scorching or its buds from freezing. But it also needs access to nutrients that are seldom conveniently dissolved in water: nitrogen in forms plants can use, phosphorus that is often locked to mineral surfaces, micronutrients that are present but stubborn. In most forests those elements are not delivered by the root alone. They arrive by way of partnerships.

A mycorrhiza is the intimate alliance between a plant root and a fungus. The root offers the fungus sugars made by photosynthesis; the fungus offers the root a vastly expanded reach into the soil. The fungal threads (hyphae) are thinner than roots, and can explore pores that roots cannot enter, picking up nutrients and water, ferrying them back to the root surface. In many forests, these partnerships link multiple plants into what ecologists call a common mycorrhizal network: shared fungal pathways that connect neighbors, sometimes of different species. The details vary from place to place—some forests are dominated by ectomycorrhizal fungi that sheath roots, others by arbuscular mycorrhizal fungi that enter root cells—but the consequence is similar. The ground is not merely a medium; it is an infrastructure.

A clear-cut does not merely remove the trees; it also removes the energy source for that infrastructure. Fungi are not green. They cannot photosynthesize. Their sprawling subterranean bodies are built and maintained with carbon paid for by living plants. When the canopy is taken away in one operation, the steady flow of carbon to the fungal network collapses. Some fungi can persist on stored reserves for a time; some can switch to decomposing dead wood; some form resistant spores. But the living, functioning network that was tuned to that particular community of trees begins to thin, fragment, and in many patches wink out. That loss is often described in chemical terms: fewer mycorrhizae means poorer nutrient uptake. It is also a physical loss. Fungal hyphae help bind soil into stable aggregates, the crumbly architecture that makes pores for air and channels for water. Without them, soil structure can slump. Rain can hit bare ground and seal the surface. Water runs off instead of infiltrating. A seedling planted into that can survive, but it is asked to do so with one hand tied behind its back.

Foresters have long noticed an old, practical truth: seedlings do better when they are not alone. The classic image is of natural regeneration under scattered "seed trees," an approach used for centuries in various forms. The modern retention idea pushes that intuition further, with a focus not only on seeds raining down from survivors, but on the below-ground continuity those survivors maintain. The survivors function as *lifeboats*—not in the sentimental sense of "saving the forest," but in the technical sense of carrying living communities through a bottleneck.

One of the most vivid demonstrations comes from the Pacific Northwest, where clear-cutting became an emblem of industrial forestry in the twentieth century. By the 1980s and 1990s, a combination of ecological research and public pressure began to change the conversation about what a harvested stand should look like afterward. Franklin and colleagues, working in this region, helped popularize variable retention as a way to preserve biological legacies: live trees, dead standing snags, down logs, patches of intact forest. The point was not merely to leave a few "pretty" trees for the view from the road. It was to keep functioning pieces of the old system present while the new system assembled itself.

A retained tree is not simply a pole. It is a habitat tower above ground, but it is also a hub below ground. An older tree has a larger root system, and with it a larger mycorrhizal community. Its roots have been negotiating with fungi for decades, selecting partners, tolerating some, excluding others, and creating a local history written in spores and hyphae. When that tree remains after harvest, its fungal partners retain a source of carbon, and the network can persist in the soil around it. This matters most in the first harsh summers after logging, when the soil is hotter and drier and more exposed than it used to be. The seedlings that establish within reach of that retained root-fungus complex are, in a real physiological sense, not starting from scratch. They are plugging into a surviving circuit.

The arrangement of retention changes the shape of that circuit. Leaving an even sprinkling of isolated trees can help, but the trees may be stressed by wind exposure, heat, and altered water dynamics. Leaving small patches, by contrast, can maintain something closer to an intact microclimate: shaded soil, moderated temperatures, and a reservoir of roots that remain intertwined. Patches also protect each other from the physics of disturbance. A lone tree in an opening becomes a sail in a storm; a small group shares the load. The geometry above ground therefore influences the continuity below ground.

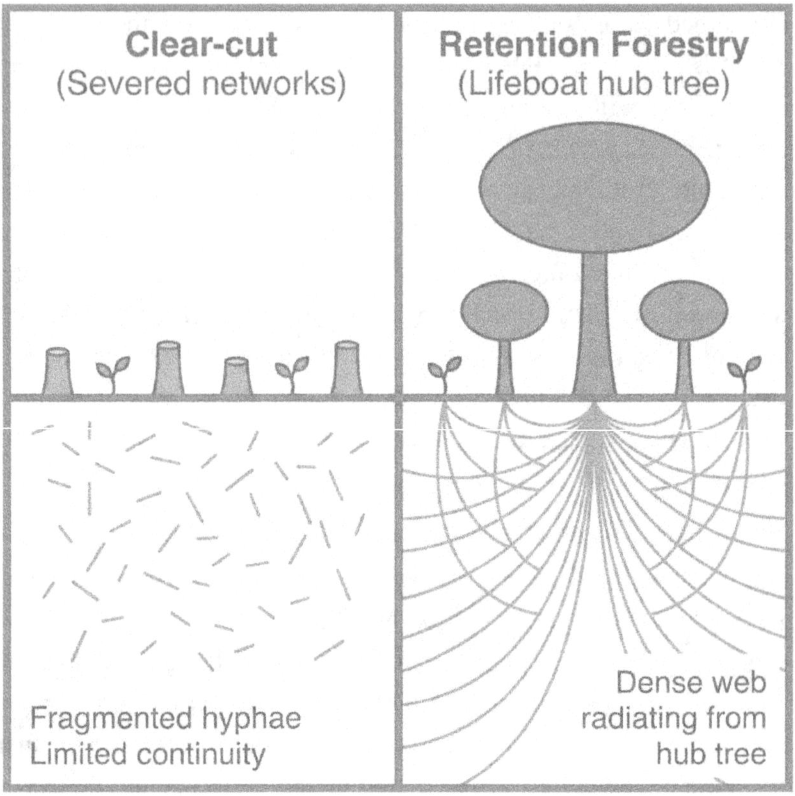

The idea of a "hub tree" can be misunderstood as mysticism: a single wise elder tree dispensing help to the young. What makes it work is more ordinary and more compelling. It is the continuity of carbon and the continuity of contact. Fungi that were already present remain fed, so their hyphae remain alive, and the young roots that arrive later meet partners that are already organized into a working mesh. The retained tree also carries a bank of fungal diversity. Different mycorrhizal fungi have different talents: some are better at scavenging phosphorus, some at tolerating drought, some at navigating particular soil chemistries. Diversity matters because the first years after harvest are unpredictable. A warm spring can push budbreak early, then a late frost can kill tender shoots; a summer can swing from saturated to desiccated; a disease can flare in the stressed microclimate. A broader fungal toolkit increases the chance that some partnerships will fit the moment.

There is also a quiet, physical engineering role that retention protects. Soil is often spoken of as if it were simply ground-up rock plus decayed leaves. In reality, a fertile forest soil has architecture: clusters of mineral grains glued into aggregates, separated by pores that hold air and water, threaded by roots and hyphae. This structure is not merely tidy; it is protective. When organic matter is tucked inside aggregates, it becomes harder for microbes to consume quickly. Carbon lingers longer. Water infiltrates rather than skating across the surface. During storms, aggregates resist being broken apart and washed away. Fungal hyphae contribute to this architecture by physically enmeshing particles and by producing sticky compounds and residues that help bind soil. When the fungal network collapses after clear-cutting, it is not only nutrient foraging that weakens. The soil's capacity to remain porous and cohesive can weaken too, leaving young plants exposed to drought and erosion as much as to hunger.

Retention does not eliminate disturbance; it redesigns it. A harvested stand is still brighter, windier, and noisier than before. But it can remain biologically inhabited in ways that matter. Consider what happens to a seedling planted into a clear-cut, nursery-grown in sterilized substrate. Its roots may carry few compatible fungal partners. In the best case, spores in the soil colonize it quickly. In the worst case, it sits for a long time with an underpowered root system, absorbing what it can from a soil whose nutrient economy was built for an older forest. Retention changes the odds. The survivors and patches function like inoculation points scattered through the site. Spores are present, yes, but so are active hyphae and colonized roots. For many mycorrhizal fungi, living contact is faster and more reliable than waiting for a spore to germinate and locate the correct host in stressed soil.

There is a human story embedded here, because retention forestry did not arise only from laboratory findings about fungi. It also arose from the friction between industrial efficiency and lived landscapes. In regions where clear-cuts bordered towns, rivers, and recreation areas, the abruptness of the practice was not just visually shocking; it made people feel that something continuous had been broken. The early debates about "leave trees" were often framed around scenery and wildlife, the things most people can see. The underground networks were not part of public language. Yet the policies that resulted—buffers along streams, retention patches, the protection of legacy trees—ended up preserving the invisible systems as well. Sometimes environmental reforms succeed not because everyone agrees on the same reasons, but because the chosen actions protect multiple values at once. Leaving living structure after harvest keeps birds, keeps shade, keeps beauty, and also keeps the fungal infrastructure that makes reforestation more than a planting campaign.

Retention also changes the meaning of dead wood. Clear-cut sites often appear littered with debris, and debris can look like waste. In a networked forest, down logs and stumps are not merely leftovers; they are bridges. Decaying wood holds moisture, moderates temperature, and provides a substrate for fungi that specialize in wood rather than roots. Those decomposers release nutrients slowly back into the soil, and their own hyphae can intersect with mycorrhizal networks in the crowded underground. Nurse logs—fallen trunks that become cradles for seedlings—work partly because they are elevated and moist, but also because they are biologically active, seeded with fungi that can meet a young root. Retention practices that keep some large woody debris on site therefore preserve not only habitat but also the microbial tempo of decomposition and nutrient release.

None of this suggests that retention is a simple "leave a few trees" recipe. The wrong trees, left in the wrong pattern, can fail. Windthrow can turn retained stems into snapped spears. Sun exposure can stress shade-adapted species. Retained patches can become too small to buffer temperature swings, or too isolated to share fungal diversity across the site. Even when the approach works, it can feel slower and messier than the clean geometry of a clear-cut followed by uniform replanting. But forests are not built by cleanliness. They are built by continuity.

A useful way to think about retention is to imagine the forest as a city whose roads, pipes, and power lines are underground. A clear-cut is not merely the demolition of buildings; it is the tearing up of utilities. The new buildings can go up quickly, but life inside them stutters until the infrastructure returns. Retention leaves enough of the old grid in place that the new growth can connect to it. The seedlings still have to grow, and storms will still come, and drought will still test them. Yet they do not face those tests as solitary organisms, each probing outward in ignorance. They begin life already surrounded by living filaments that know the soil's hidden corridors, already within reach of a web that can hold moisture a little longer, trade nutrients a little farther, and keep the ground itself from unravelling into dust.

9.2 Inoculating the Future

A sapling raised in a nursery is, in one sense, a triumph of care: regular water, protection from browsing deer, measured light, freedom from competition. In another sense it is an exile. Many tree seedlings are cultivated in potting mixes designed to be clean, uniform, and predictable. Cleanliness keeps damping-off fungi at bay and makes growth more controllable. Uniformity simplifies schedules and budgets. Predictability is what large planting programs are built on. Yet a forest is not predictable, and a sterile childhood can be a poor preparation for life in wild soil.

The shock arrives at planting. A seedling's roots meet a world of mineral surfaces, decaying litter, acidic pockets, droughty ridges, saturated hollows, and microbes already fighting over every molecule of carbon they can reach. In that world, the seedling's most important allies are often not visible above ground. They are the fungi that form mycorrhizae: partnerships in which roots trade sugars for assistance in finding water and nutrients, and in coping with stress. A tree can live without these partners, but it will usually be smaller, slower, and more vulnerable. In a degraded site, or on a fresh clear-cut where the old networks have been damaged, "usually" can become "often fails."

Soil inoculation is a simple-sounding intervention with surprisingly deep implications. It means reintroducing beneficial microbes—especially native mycorrhizal fungi—to the roots of planted seedlings or to the soil around them, so that the partnership is established early rather than left to chance. The goal is not to sprinkle magic dust. It is to restore a missing piece of ecology that industrial practices often remove.

The need becomes clearer if one keeps in mind what mycorrhizal fungi actually do for a tree. Roots are not straws; they are negotiating surfaces. Nutrients such as phosphorus, in particular, are frequently present in soils but held tightly to minerals or bound in organic matter that plants cannot directly access. Mycorrhizal hyphae are exquisitely good at exploring tiny pores and deploying the right chemistry in the right place, pulling scarce nutrients into the living economy of the root. Nitrogen, too, is often inaccessible in its most abundant forms, and is made plant-available through a chain of microbial transformations. Fungi and their microbial associates can sit at key points in that chain, increasing the rate at which a seedling gains usable nitrogen relative to its neighbors. The fungus also helps with water, not by acting as a pipe so much as by extending the functional volume of soil the plant can draw on, smoothing the difference between a wet week and a dry one.

The benefits are not only chemical. Fungal hyphae also help build soil structure. Soil that is well aggregated has a kind of crumbly coherence; it resists crusting, allows infiltration, and holds both oxygen and water in the right proportions. Hyphae physically entangle particles and contribute sticky residues that help aggregates persist. That structure matters to a seedling because it determines whether the soil behaves like a sponge or like a brick. A brick forces roots to work harder for every millimetre of penetration and makes drought arrive earlier.

All of this makes mycorrhizal partnerships sound universally good, and in broad outline they are. But inoculation introduces a complication that is both scientific and practical: *which* fungi? Forests do not have a single "mycorrhiza," any more than cities have a single "transport system." They have suites of fungi with different strategies, and those strategies interact with the identity of the host plant, the chemistry of the soil, the climate of the site, and the history of disturbance. Inoculation works best when it is not an imported product but a thoughtful matchmaking.

The idea of deliberately giving plants microbial partners is not new. By the late nineteenth century, agronomists already knew that legumes benefited from nodules filled with nitrogen-fixing bacteria, and experiments began to show that adding the right bacteria could improve crop performance. The language of "inoculation" became familiar in agriculture: coat seeds with a culture, plant into poor soil, and watch the crop turn greener. Mycorrhizal fungi were harder to harness, partly because many cannot be grown like common bacteria in a flask, and partly because their life cycles are entwined with living roots. Still, foresters and horticulturalists noticed early that some seedlings transplanted from wild soil seemed to establish with unusual vigor, while others languished despite identical watering and care. The difference was often in the roots, invisible unless one bothered to look closely.

The modern push for inoculation in restoration grew in part from a sobering pattern: large-scale reforestation projects can meet their planting targets and still fail ecologically. Survival rates can be low, growth can stall, and planted trees can remain physiologically stressed for years. Managers often reach first for explanations they can see—insufficient rainfall, browsing, competition from grasses, poor seed stock. Those are real. But underneath them lies another failure mode: the planted seedlings never acquire the right symbionts, or acquire them too late, after the most critical early years have passed.

A seedling's first year is a bottleneck. Its root system is small, its carbohydrate budget is tight, and its ability to buffer temperature and moisture extremes is limited. If the seedling must also "audition" for fungal partners in soil where those partners are sparse or absent, it may lose the race before the partnership begins. Inoculation aims to remove that particular bottleneck by arriving at planting already allied.

How is inoculation actually done? The methods range from the rough-and-ready to the carefully engineered. At one end is *soil transfer*: bringing a small amount of soil or forest litter from a healthy donor site and mixing it into planting holes or nursery beds. This can introduce a whole community: spores, hyphae, bacteria, microfauna, and the chemical signatures that tell roots what kind of neighborhood they have entered. Soil transfer can be powerful because it does not require a manager to pick the "correct" fungus; the community arrives as a package that has already coexisted with local plants. It also carries risks: pathogens can hitchhike, and removing soil from intact habitats can be ecologically costly unless done with care.

At the other end are *cultured inocula*: preparations produced in controlled conditions, sometimes containing a few selected mycorrhizal species, sometimes designed for a particular plant group. These can be applied as a slurry to roots, incorporated into potting mixes, or placed in pellets near roots at planting. Cultured inocula are easier to scale and easier to standardize. Their weakness is that they can become a one-size-fits-all solution for a world that is not one-size-fits-all. A fungus that thrives in a greenhouse mix and colonizes quickly may not be the fungus that best supports a tree under drought on a nutrient-poor slope.

Between these extremes are hybrid approaches that have become increasingly common in ecological restoration: "native inocula" produced from local soils, often multiplied using trap cultures with host plants, then introduced back into restoration sites. The promise is local adaptation without stripping donor sites. The practical challenge is time, expertise, and the unavoidable messiness of living communities.

Even when inoculation is successful, it rarely shows its full value immediately. A fungus is not a fertilizer that instantly dissolves. It has to colonize roots, expand hyphae into soil, build a network, and then translate that network into improved nutrient and water access. Across multiple field studies, the strongest benefits often emerge over years rather than weeks. That time lag matters because it changes how success should be measured. A one-season trial can miss what an inoculum is quietly building. In long-term restoration experiments, inoculated plots can diverge from uninoculated ones gradually, then decisively, as plants pass through repeated stresses and the inoculated individuals persist, grow, and reproduce more effectively.

The ecological consequences extend beyond the target trees. Mycorrhizal fungi can influence which understory plants establish, how quickly weeds invade, and how plant communities assemble. When a restoration effort is described as "planting trees," it is easy to forget that a forest is also a botanical neighborhood of shrubs, herbs, mosses, and the animals and microbes that track them. Introducing mycorrhizal partners can shift that neighborhood. Certain fungi favor certain plant functional types; some partnerships are particularly beneficial to nitrogen-fixing woody plants, for example, because the demand for phosphorus rises when nitrogen becomes abundant. In nutrient-poor soils, these interactions can determine whether a site becomes a stable, diverse community or a plantation that is perpetually on the edge of collapse.

There is also a moral dimension to inoculation, not in the sense of good and evil, but in the sense of responsibility. When people disturb a landscape at industrial scale—logging, mining, construction, agriculture—they often remove the "slow" parts of ecology. Plants can be replanted quickly. Microbial communities and soil structure rebuild slowly. A restoration project that ignores microbes is, in effect, asking plants to rebuild the slow parts alone. Inoculation is an admission that the work must be shared: humans replant, and also help return the invisible partners that make replanting meaningful.

Yet inoculation can be oversold, and it is worth being honest about its limits. Not every site needs it. In soils that retain intact mycorrhizal communities, inoculation can do little or nothing because the local fungi colonize roots readily anyway. In some cases added fungi simply fail to establish, outcompeted by resident communities or mismatched to the soil's pH and nutrient profile. In other cases the inoculum establishes but brings a partnership that is not optimal under local conditions. Mycorrhizal relationships are mutualistic, but they are also economic. A fungus can take carbon from a plant while providing relatively little in return if nutrients are abundant or if the fungal species is poorly matched. The plant may tolerate this when it can afford to, but a newly planted seedling often cannot.

This is why the best inoculation strategy looks less like a product and more like a diagnosis. What is the site history? Was the soil stockpiled, compacted, or eroded? Has topsoil been removed? Is the site low in available phosphorus? Are there remnant patches nearby that can naturally supply spores and hyphae? Are the planted species ones that strongly depend on mycorrhizae? The answers shape the decision. The most compelling cases for inoculation are those where disturbance has been severe enough to deplete the local fungal pool, and where seedlings face a gauntlet of stress in their first years.

A concrete example makes this easier to picture. Imagine a hillside logged repeatedly and then burned, with machinery compacting the soil and rain washing fines downslope. The surface becomes a patchwork: exposed mineral soil in places, dense grass in others, and pockets of surviving shrubs. Planting conifers into such a site can yield disappointing results: seedlings survive but remain stunted, yellowish, and prone to drought stress. Inoculating those seedlings with native mycorrhizal fungi at planting can change the trajectory not because it "adds nutrients" but because it restores access. Hyphae extend beyond the compacted rhizosphere, reaching pockets of moisture and nutrient-rich microsites. Over time the fungal network contributes to aggregation, improving infiltration so that water from winter rains is held longer into summer. The seedlings do not become immune to drought, but they gain a margin—often the margin between persistence and failure.

A historical anecdote from the early era of intensive plantation forestry captures the intuition behind this, even before mycorrhizae were part of everyday forestry language. In parts of Europe where forests were converted to uniform stands and soils were heavily disturbed, foresters sometimes observed that seedlings planted near remnant woodland edges or near older trees established more vigorously than those planted in the middle of large open areas. The edge offered shade and wind protection, but it also offered something subterranean: a living soil community extending outward from intact forest. Managers who retained strips, patches, or scattered veterans were sometimes motivated by aesthetics or hunting, yet they inadvertently preserved microbial continuity. What looked like sentiment could function as infrastructure.

Inoculation also forces a reconsideration of what "native" means in a changing climate. Restoration often aims to recreate a reference ecosystem, a snapshot of what grew there before disturbance. But climate is no longer stable. Droughts intensify in many regions, and heat waves arrive at odd times of year. A seedling planted for "historical fidelity" may now face conditions outside the envelope its local fungal partners are adapted to. Conversely, fungi adapted to drier, warmer conditions may not be present locally in sufficient abundance. This raises difficult questions: should inoculation always be strictly local, or can it sometimes be regional, selecting fungi from environments that resemble the site's likely future? The question is uncomfortable because moving microbes can also move pests and pathogens, and because ecological interventions have a history of unintended consequences. Still, it is hard to avoid the logic: if a tree is being asked to live in a future climate, its microbial partners may need to be chosen with that future in mind.

The practical reality is that inoculation is most effective when treated as part of a broader design, rather than a stand-alone fix. Retention forestry preserves lifeboat networks on site; inoculation rebuilds alliances where those lifeboats were lost. Together they shift reforestation from an engineering project to a kind of assisted reassembly. The visible work is still done with shovels, planting bars, and trucks.

The lasting work is done in millimetres, as hyphae spread through pores, as roots respond by branching more finely, as aggregates form and persist, as nutrients once locked away begin to circulate.

Perhaps the most hopeful aspect of inoculation is that it acknowledges the intelligence of the soil without romanticizing it. A forest floor is not a passive substrate waiting to be planted; it is a living membrane that organizes water, carbon, and minerals into patterns that plants can use. When those patterns have been disrupted, reintroducing a few grams of the right biology to a seedling's roots can be more consequential than adding kilograms of fertilizer. The fertilizer is spent quickly; the partnership can compound, year after year, turning survival into growth and growth into the return of shade, litter, and the slow rebuilding of a habitat that feels, from above, like a forest again.

9.3 Migrating the Microbiome

The phrase *assisted migration* has the clipped sound of a bureaucratic memo, but it points to a problem that is intimate and immediate: trees cannot walk away from danger. A mature pine or fir is anchored to a particular hillside, and even a sapling can only move as far as its seed can travel. Climate, meanwhile, is moving with the wind. In many parts of the world, the long-term averages that shaped local forests are shifting poleward and uphill, while extremes—heat waves, drought, sudden cold snaps after warm spells—arrive like ambushes. A forest can endure change; it evolved with fire, storms, and insect outbreaks. What it struggles with is the loss of timing and the compression of options, the way the future can arrive faster than a generation of trees can respond.

For the last decade or two, the public conversation about forest adaptation has tended to focus on the visible actors: which tree species to plant, which seed sources to choose, whether to move genotypes from warmer regions into cooler ones. These decisions are real and controversial. They raise ethical questions about introducing organisms beyond their historical ranges and practical questions about what will survive. Yet trees are not solitary units that can be shipped and installed like lampposts. Every tree is a bundle of relationships, and many of those relationships live in the soil. When people talk about moving trees without talking about moving their fungi, they are imagining an organism without its organs.

A mycorrhizal partnership is often described as a mutualism between a root and a fungus, but in practice it is also a cultural inheritance. Many mycorrhizal fungi disperse slowly. Their spores are heavier than pollen and often travel by animal vectors or by local wind currents that do not carry them far. Their best dispersal mechanism can be continuity: spreading through existing networks and colonizing new roots nearby. When climate zones shift, seeds can sometimes leap ahead on the wind or in the stomach of a bird. Fungi frequently lag behind. A seed arriving in newly suitable habitat may find that the air temperature is right and the rainfall pattern is acceptable, but the below-ground partners it needs are absent or rare. It is like relocating a person to a city where the buildings are ready but the water system is missing.

This "lag" is not merely a curiosity. It can become a bottleneck that determines whether a forest can expand into new territory fast enough to keep pace with climate change. A tree species that depends strongly on particular mycorrhizal partners may be constrained by the dispersal of those partners. Even when fungi are present, the local community may not include the strains best suited to the tree's physiology or to the new site's soil chemistry. The gap can be felt most strongly at the seedling stage, when a plant's own foraging capacity is smallest and when drought or nutrient limitation can kill quickly.

The idea of migrating the microbiome therefore follows naturally from retention and inoculation. Retention preserves living underground structure where disturbance would otherwise erase it. Inoculation restores missing partners to seedlings planted into damaged soils. Migrating the microbiome takes the logic into the future tense: if forests are being asked to shift their ranges, the relationships that make them work need to shift too.

It is tempting to imagine that this happens automatically. After all, fungi have been moving with plants for hundreds of millions of years. The trouble is that natural migration happened in a world where climate zones moved slowly and where landscapes were not chopped into roads, farms, suburbs, and fire-suppressed fuel loads. After the last glaciation, trees expanded northward over millennia; fungal partners had time to follow. Today, suitable climate for some species is projected to move hundreds of kilometres in a century, while the landscape is fragmented and disturbances are amplified. Natural dispersal is still at work, but it may not be fast enough.

There is a historical analogy that makes the problem legible. In the late nineteenth and early twentieth centuries, European colonial powers moved plants around the world with extraordinary speed: rubber trees, tea, cacao, cinchona. Sometimes the plants flourished. Sometimes they failed for reasons that were mysterious until later: absence of pollinators, presence of novel pests, or incompatibility with local soils. Modern forestry has its own history of ambitious planting programs that underperformed, not because the trees could not grow in the climate, but because the ecological context was wrong. The microbiome is part of that context. Moving trees without their microbial partners can reproduce the old pattern: impressive effort, disappointing resilience.

So how might one move a microbiome responsibly? The simplest strategy is to move a little soil along with the plant. A handful of earth looks trivial, yet it can contain spores, hyphal fragments, bacteria, and the chemical cues that help symbionts establish. In restoration ecology this approach is often called a *soil transplant* or *soil transfer*. It is older than the terminology: gardeners have long known that moving some soil from a thriving bed can help a struggling one. In forest contexts, soil transfer can be done at multiple scales: from mixing donor soil into nursery pots so seedlings are colonized before planting, to placing small inoculum plugs into planting holes, to spreading thin layers of forest floor material across degraded sites.

What makes soil transfer scientifically interesting is that it can move not just species but *interaction patterns*. A microbial community is not merely a list of taxa; it is a network of competition, cooperation, predation, and chemical signaling. Different donor soils can introduce communities with different "central" organisms—microbes that, by virtue of their interactions, influence the rest of the community's structure. Experiments in which soils from different donor sites are added to the same degraded recipient site often show that the donor identity can steer the trajectory of both the microbiome and the plant community for years. The microbiome does not simply snap back to a generic local state; it can retain the signature of the donor, particularly when the transfer is large enough and when the recipient site is severely depleted.

This is both promising and unsettling. It suggests that microbiomes can be deliberately moved and can persist, making it feasible to pair trees with their preferred fungal allies when establishing forests in new climatic zones. It also suggests that the choice of donor matters enormously. A donor soil is not an inert ingredient; it is an ecological decision with long consequences.

The scale question is unavoidable. Moving meaningful quantities of forest soil is expensive and ecologically fraught if it damages donor sites. The solution, when possible, is multiplication: using small donor samples to start cultures that can be expanded without continuously mining intact forests. For arbuscular mycorrhizal fungi, which cannot be easily grown without a host plant, "trap cultures" can be used: grow suitable host plants in a controlled medium seeded with a small amount of donor soil, allow the fungi to colonize and produce spores, then harvest the inoculum. For ectomycorrhizal fungi, some species can be cultured more directly, while others remain difficult. Even with multiplication, however, the inoculum remains a simplified version of a natural forest floor. It can be helpful, but it is not a full transplant of a mature ecosystem.

A concrete contemporary example helps. Consider a reforestation project in a mountainous region where warming is pushing suitable habitat uphill. Managers may decide to plant seedlings sourced from slightly warmer, lower-elevation populations, betting that these genotypes are pre-adapted to future temperatures. If those seedlings are planted into high-elevation soils that have never hosted that particular tree species, they may encounter a mismatch: the local mycorrhizal community may not include the fungi that most effectively partner with the incoming tree genotype. Growth may be slow, and drought years may cause high mortality. Introducing a small amount of soil or inoculum derived from the tree's original range can change the outcome by ensuring that seedlings form functional mycorrhizae early. Over multiple seasons, the transplanted fungal community can remain a major determinant of which fungi colonize the roots, even as the surrounding soil microbes influence the final composition. The seedlings, in effect, arrive with a familiar support team.

The phrase "remain a major determinant" is doing a lot of work. Persistence does not mean dominance everywhere, nor does it mean the inoculum will function as intended. Environmental conditions can block establishment. Soil pH, temperature regime, moisture pattern, and nutrient availability can all influence whether a transplanted fungal community colonizes roots robustly. A fungus that thrives in one environment can be present but functionally marginal in another. This is the compatibility constraint that keeps microbiome migration from becoming a simplistic panacea. Even if the introduced fungi survive, they may not deliver the same benefits if the new soil chemistry changes the cost-benefit balance of the partnership.

This is where the idea of moving "the microbiome" becomes more than a technical matter; it becomes a conversation about what forests are allowed to become. A strict historical view would insist that only local fungi should be used, because they are part of the native ecosystem. A purely utilitarian view might argue for importing whatever fungi make trees grow fastest. Neither position is adequate in a world where "local" is itself changing. Climate change is not merely shifting temperature; it is reorganizing fire regimes, drought patterns, insect pressures, and the timing of snowmelt. A forest that remains in place may nonetheless become climatically foreign to itself. Under those conditions, insisting on strict local provenance for microbes can be as artificial as importing a commercial mix from the other side of the planet.

Responsible microbiome migration therefore resembles careful plant sourcing: regional, not global; targeted, not indiscriminate; paired with monitoring, not done as a one-off intervention and forgotten. The most cautious approaches look for donor sites that resemble the recipient site's likely future conditions—similar soils, similar seasonal moisture patterns, similar disturbance regimes—and then move small inoculum quantities that can establish without overwhelming local communities. This is not a guarantee against unintended consequences, but it is a way to reduce the probability of introducing organisms that become pests or that disrupt existing mutualisms.

Pathogens are the shadow that follows every discussion of moving soil. Forest pathology offers too many cautionary tales: chestnut blight, Dutch elm disease, sudden oak death. Most of these catastrophes are not about mycorrhizal fungi, but they teach the same lesson: organisms travel easily when humans move them, and ecosystems can be fragile to novel enemies. Soil transfer can move not only beneficial fungi but also harmful oomycetes, nematodes, or other pathogens. This is one reason why inoculum production is increasingly paired with screening, quarantine practices, and an emphasis on using local or regionally appropriate sources. It is also why some practitioners prefer cultured inocula that are more controlled, even if they are less diverse.

There is another, subtler risk: moving a microbiome can change the competitive balance among plants. Mycorrhizal fungi can favor some species over others, and a shifted fungal community can tilt a restoration toward certain outcomes. This can be desired when invasive weeds dominate and native plants struggle; a transplanted native mycorrhizal community can help natives establish and suppress invaders. But it can also produce unexpected winners, especially in communities where multiple plant species are being introduced together. A reforestation project is rarely just trees; it is also the understory that will eventually define fire behavior, wildlife habitat, and water cycling. Shifting the microbiome is, in that sense, shifting the future forest's social order.

What makes all this worthwhile is that the alternative is not neutrality. Doing nothing is also an intervention when the climate is changing rapidly. A seedling planted into a new range without compatible fungal partners is not being left "natural"; it is being placed into an artificially simplified context created by disturbance, fragmentation, and accelerated climate change. The question is not whether to intervene, but how to intervene with humility.

It is easy to picture microbiome migration as an exotic new frontier, but in a quieter way it is already happening. Every time topsoil is moved by construction equipment, every time potted plants are shipped across regions, every time wildfire suppression changes which fungi dominate by changing which plants dominate, microbial communities are being rearranged. The difference is that these rearrangements are usually accidental. The emerging proposal is to make them intentional, evidence-based, and oriented toward ecological function rather than convenience.

The most persuasive argument for moving fungi with trees is not philosophical but practical: the early years of a planted forest are when mortality is highest and when the trajectory of the whole community is set. If the goal is not merely to plant stems but to establish self-sustaining forests that can hold soil, store carbon, and provide habitat under a harsher climate, then the alliances that allow trees to feed, drink, and tolerate stress deserve as much attention as the seedlings themselves. Moving a microbiome is an attempt to carry the living memory of a functioning forest into a place where a functioning forest must be built quickly.

One can imagine a future planting crew whose equipment includes not only seedlings and protective tubes, but small, carefully sourced packets of soil life: microbial consortia matched to soil types, climates, and host species, tracked like seed lots and monitored over time. The scene would look almost the same from the road: people with shovels on a hillside. The difference would be in the intention. Each tree would be planted not as an isolated organism expected to improvise its own support system, but as the visible tip of a community designed to take root.

Forests have always depended on migration. After ice sheets retreated, trees followed the warming northward; after fires, new cohorts followed the ash. The difference now is the speed, and the fact that humans have become the main vector for both disturbance and repair. If the repair is to be more than cosmetic, it will have to include the parts of forests that do not pose for photographs: the spores on a boot sole, the hyphae threaded through a crumb of soil, the microbial conversations that determine whether a seedling's first summer ends in growth or in desiccation.

Chapter 10

Beyond the Hype

The 'wood-wide web' has captured the popular imagination, yet the scientific reality is far more nuanced-and debated-than the headlines suggest. By stripping away anthropomorphic metaphors of forest socialism, we reveal a biological architecture that is not necessarily altruistic but is astonishingly robust, offering profound lessons for human engineering and the future of biosphere stewardship.

10.1 The Altruism Illusion

A good story travels faster than a careful one. Few ecological stories have travelled faster, in recent years, than the idea of the *mother tree*: an old, wise giant in the forest, connected to her kin through a lacework of fungal threads, sending food and warnings to needy seedlings the way a parent slips lunch money into a child's pocket. It is a story with a moral spine. It reassures us that nature, left to itself, tends toward care; that the living world is not only competition but community; that the forest is, in a quietly literal sense, *connected.*

The trouble is that moral clarity is not the same thing as biological clarity.

Start with the part almost everyone agrees on. Many plants form mycorrhizas: partnerships between roots and fungi. The fungus extends the plant's reach into the soil with fine filaments called *hyphae*;

the plant pays, largely in sugars and fats made by photosynthesis. In many cases, those hyphae also connect the roots of different plants. When the same fungal individual (or genetically identical network) links more than one plant, ecologists call it a *common mycorrhizal network* (often shortened to CMN). If the word *network* evokes fibre-optic cables and deliberate routing, it can mislead; a CMN is not a designed grid but an entangled growth form shaped by opportunity, physics, and survival. Still, the geometry matters. A fungal web can, in principle, move resources from one place to another.

Where disagreement begins is not with *whether* resources can move, but with what the movement *means*.

A concrete example anchors the modern debate. In the late 1990s, Suzanne Simard and colleagues performed a field experiment that became a touchstone for the "tree-to-tree transfer" idea. Using isotopic tracers—essentially tagging carbon so it could be followed—they showed that carbon could move between different tree species (paper birch and Douglas-fir) in a forest setting, and that the direction and net outcome depended on conditions such as shading. This was exciting for a simple reason: it offered a mechanism by which one plant's photosynthesis might partly subsidize another's. The result was not a fairy tale; it was a measured, conditional transfer in a particular ecological context. But in the retelling, conditionality tends to evaporate. The experiment becomes a parable: birch "feeds" fir; elders "nurse" the young.

If you want a historical parallel, consider how the earliest descriptions of coral reefs read to European naturalists. Reefs were portrayed as bustling cities of cooperation, harmonious architectures built by tiny animals. Only later did the harsher realities come into view: intense competition for light, chemical warfare between neighbours, opportunistic predation, parasitism, disease. The city metaphor was not wrong, exactly, but it encouraged a one-sided attention. Something similar happens when a forest is cast as a family.

To see why, it helps to ask a blunt question: if a seedling receives carbon that once belonged to an adult tree, is that a 'gift'?

Not necessarily. In biology, 'gift' implies a cost borne by the giver that is not repaid, and that increases the recipient's fitness. Demonstrating that chain requires more than detecting a labelled molecule in a neighbouring plant. A carbon atom can move for many reasons. It can leak from roots as exudates and be taken up indirectly. It can move through soil water as dissolved organic carbon. It can be respired by microbes and re-fixed later by photosynthesis. Even if the pathway is truly fungal, a transfer can be a side effect of fungal transport and storage rather than an intentional allocation from one plant to another.

The word *intentional* is dangerous here, because neither trees nor fungi need intentions for their behaviour to look purposeful. Local rules—grow where sugars are abundant, allocate more hyphae to richer patches, thicken high-flow cords—can generate system-level patterns that resemble planning. Humans are exquisitely tuned to mistake such patterns for motives, the way we see faces in clouds or meaning in coincidences. The "mother tree" story is persuasive partly because it offers a motive we can recognize.

A more cautious reading of the evidence has been gaining strength, especially as researchers have stepped back to evaluate the field as a whole rather than highlight its most vivid individual studies. Justine Karst and colleagues, writing in 2023, made an uncomfortable argument: that popular claims about CMNs in forests have run ahead of what the data can firmly support, helped along by positive citation bias and a habit of overinterpreting results. "Positive citation bias" sounds technical, but the idea is familiar from everyday life. If people keep repeating the most dramatic version of a story and quietly forget the caveats, you end up with certainty built out of emphasis rather than proof.

The central claims under scrutiny are not trivial. They include the idea that CMNs are ubiquitous in forests; that they routinely move resources in ways that improve seedling performance; and that mature trees preferentially send resources or signals to their own offspring. Each of these claims requires distinct kinds of evidence. Mapping a fungal network is not the same as showing that it transfers carbon. Showing a transfer is not the same as showing a growth benefit. Showing a growth benefit is not the same as showing nepotism.

Even defining "connected" is harder than it seems. A seedling's roots may carry fungal structures, but that does not guarantee that the fungus is the same individual linking it to a neighbouring tree. Many fungi have complex life histories; what looks like one continuous mat may be a patchwork of genetically distinct individuals, each with its own 'interests.' In dense soils, hyphae from different fungi overlap and fuse and fragment. Some fungi can anastomose—fuse hyphae together—but fusion is not a universal handshake; compatibility matters, and even compatible fusions do not guarantee free exchange. If you cannot identify the fungal individual, you may be attributing transfer to a "network" that is actually a crowd.

Then there is the measurement problem. Isotopic labelling is powerful, but it is not a magic arrow pointing to one route. When a labelled carbon signature appears in a plant, the signature testifies that carbon moved through *some* pathway, not automatically that it travelled directly from one plant to another through hyphae. To nail the CMN route, researchers may trench the soil to sever hyphae, use mesh barriers that allow fungal filaments but exclude roots, or compare treatments with and without fungal connections. Each method has trade-offs. Trenching changes moisture and root behaviour. Mesh changes microbial communities. Severing hyphae can stress fungi and plants in ways that alter exudation. The forest is not a laboratory bench; it is a system that responds to being tested.

Suppose, however, that you *do* demonstrate fungal transfer. You still have to confront a subtle point: a forest is not a charity; it is a market of constrained choices. Carbon is not simply "food." For a tree, carbon is both energy and construction material. For a fungus, carbon is the fuel that builds hyphae, spores, and the biochemistry needed to pry nutrients from minerals. If a fungus moves carbon compounds around its network, it may be doing what any organism with a transport system does: balancing supply and demand across a body spread out in space. A labelled molecule arriving at a seedling may be less like a parent's gift and more like a truck passing through town on the way to a warehouse.

This is where the most productive alternative to the altruism narrative has emerged: treat mycorrhizal exchange as a form of trade, with bargaining, asymmetry, and the possibility of cheating. Ecologists have borrowed the phrase "biological market" to describe interactions where partners can, to some degree, choose among potential associates and adjust their investments in response to returns. The metaphor is not perfect—Toby Kiers and Stuart West argued in 2016 that market talk can drift into vagueness if it is not tied to precise mechanisms and testable predictions—but it has one major virtue. It keeps the focus on incentives rather than sentiment.

Evidence for incentive-driven exchange is not just theoretical. In 2011, Kiers and colleagues reported experiments in arbuscular mycorrhizal systems (a major type of mycorrhiza common in many plants) showing that plants can allocate more carbohydrates to fungal partners that deliver more nutrients, and that fungi can bias nutrient delivery toward roots that provide more carbon. In other words, both sides can exert control. This is not parental care; it is conditional reciprocity, a feedback loop that can stabilize cooperation even among self-interested partners.

Trade, crucially, does not imply fairness. Florian Walder and Marcel van der Heijden showed in 2012 that when two plant species share a mycorrhizal network, the terms of exchange can be radically unequal, depending on the fungal species involved. One plant may pay relatively little carbon and receive most of the nutrients delivered through the network, while another may pay heavily and get little back. If you want a human analogy, it is not a family dinner where everyone shares from the same pot; it is a negotiation where some participants have leverage and others do not.

Once you see that leverage is possible, the “mother tree” tale begins to look less like a discovery and more like a projection. Old trees are often the largest sources of carbon in a stand, simply because they have the most leaves and the deepest access to light. They also host large fungal communities. If carbon moves through fungal tissues, it is unsurprising that some of it might be detected near seedlings. But the critical question is whether the seedling is being subsidized in a way that improves its survival and whether the adult tree incurs a meaningful cost that is compensated only indirectly (say, through kin selection). In many contexts, a seedling’s fate is determined less by subtle carbon subsidies than by light, water, pathogens, herbivory, and the brutal arithmetic of density. A gram of carbon is not always the limiting factor.

There is also a conceptual trap in the way the story is often told. “The forest shares resources” slides easily into “the forest is cooperative.” Yet cooperation, in the evolutionary sense, is a specific kind of behaviour: it must increase another individual’s fitness, and it must have evolved at least partly because of that effect. A behaviour can *look* cooperative and still be driven by self-serving constraints. Consider a public water fountain in a city. People drink from it. The existence of the fountain benefits strangers. But the pipes are not altruistic; they are infrastructure, maintained because the city chooses to maintain it. Fungal networks are not built by a city council. They are built by fungi for fungal reasons, and they persist when they pay.

A different line of evidence provides a useful reality check, because it shows that mycorrhizal networks *can* sustain strong, even startling, transfers—but not in the sentimental direction. Mycoheterotrophic plants, sometimes called "ghost plants," have evolved to live with little or no photosynthesis, siphoning carbon via fungi that connect them to photosynthetic hosts. If you have seen a pale, leafless plant on a dim forest floor, it may be living on borrowed carbon routed through a fungal partner. This is not a marginal curiosity. It is an entire nutritional strategy that has evolved multiple times. It strongly suggests that fungal connections can move carbon in ecologically meaningful quantities under certain conditions. Yet it is difficult to shoehorn into the mother-tree morality play, because it resembles exploitation as much as mutual aid.

The forest, then, may be neither socialist utopia nor Hobbesian brawl. It may be something more familiar and less comforting: a set of overlapping deals, some mutually beneficial, some lopsided, some opportunistic, all constrained by the physics of transport and the chemistry of soil. Hyphae are thin; moving materials through them takes energy. Nutrients are patchy; carbon is seasonal. Fungi face competition from other fungi, grazing by soil animals, drought, freezing, fire. Trees face shade, windthrow, drought, pathogens, the slow gamble of growing tall. Under those pressures, stable relationships can emerge, but stability is not the same as benevolence.

The popular imagery of the wood-wide web often depicts equal sharing: carbon drifting serenely from the strong to the weak, like warmth in a room. A more accurate picture has arrows of varying thickness, and some arrows that reverse when conditions change.

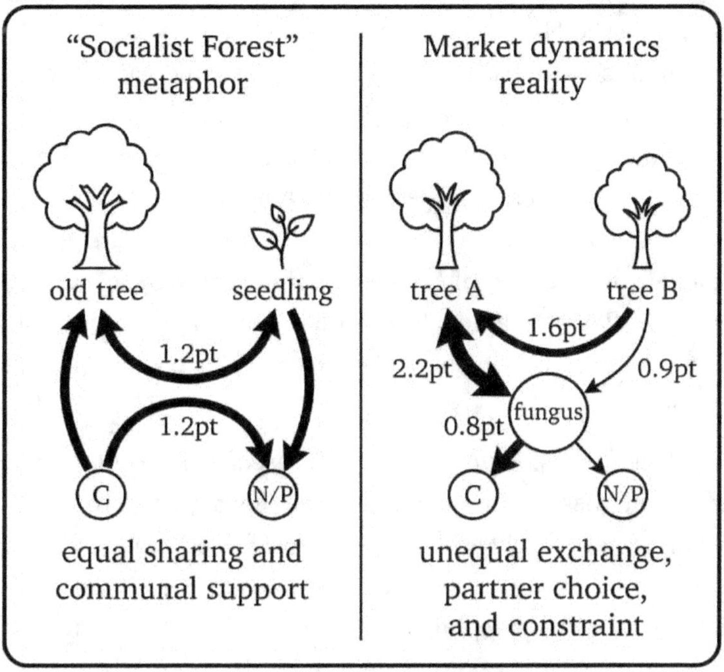

Simplified schematic: icons represent flows of carbon (C) and nutrients (N/P).

What, then, should replace the altruism headline? Not cynicism, but sharper questions.

One good question is: *who controls the valve*? If a fungus can preferentially deliver nutrients to the best-paying roots, then what looks like a tree "feeding" another tree might be a fungus reallocating its own carbon income and nutrient supply. Conversely, if a plant can reward the fungi that serve it best, then the plant is not passively embedded in a benevolent web; it is actively managing a portfolio of partners, within limits set by soil and competition. In a dense forest, a root may host multiple fungi, and a fungus may touch multiple plants. That is a crowded marketplace, not a family tree.

A second good question is: *what is the alternative pathway*? It is easy to forget that soil is not empty space around fungal wires; it is a living medium thick with bacteria, archaea, protists, microarthropods, and free enzymes. Carbon and nutrients are constantly being released, transformed, immobilized, and re-released. When an isotope label shows up somewhere unexpected, the fungus is a plausible courier, but not the only one. The most persuasive studies are those that combine multiple lines of evidence: physical barriers, genetic identification of fungal individuals, careful accounting of carbon budgets, and measurements of plant performance that extend beyond a short-term pulse.

A third question is: *what does "benefit" mean*? If a seedling receives carbon but does not grow more, survive longer, or reproduce more, then the transfer might be biologically real and ecologically minor. If it grows more in a pot experiment but not in the field, then the effect may depend on the simplified conditions of the pot. If it grows more under deep shade but not in bright light, that is still interesting—it says something about when carbon is limiting—but it is not a general claim about universal forest care.

Even nepotism, the most emotionally resonant claim, has to be treated as a demanding hypothesis rather than a default assumption. Preferential transfer to kin would require not only that transfer occurs and improves fitness, but that the system has a mechanism for kin recognition and for directing resources accordingly. Kin recognition exists in nature, but it is not free. It requires cues, discrimination, and a payoff that outweighs the risk of being deceived. In the messy soil, deception is not a hypothetical; it is the background condition.

If any of this sounds like deflation, it is worth remembering that the less romantic view is, in its own way, more astonishing. A mycorrhizal network is a living infrastructure built without a blueprint, maintained without a manager, and constantly renegotiated by organisms that do not know the forest, only the immediate chemical and physical signals at their growing tips. It can create the appearance of generosity while following rules that are closer to accounting than affection.

There is also a human reason to resist the altruism illusion. Forest management and conservation are hungry for uplifting narratives, and those narratives can become prescriptions. If we convince ourselves that old trees function primarily as benevolent donors, we may overlook the more robust reasons old growth matters: microclimate buffering, habitat complexity, seed production, pathogen dynamics, and the slow accumulation of soil structure and fungal diversity. Protecting old trees remains wise, but the wisdom does not need to rest on a single, fragile claim about parental feeding.

A forest can be connected without being kind. It can sustain transfers without promising fairness. The underground web coordinates life by moving materials through a world of scarcity, where every molecule has multiple claimants. When carbon crosses from one plant's neighbourhood to another, it is not a sermon about virtue. It is a trace of negotiations happening in the dark.

10.2 Nature's Algorithms

The temptation, when faced with a fungal network, is to treat it as a metaphor. Threads in the dark become "communication lines," forks become "choices," and a spreading mycelium becomes a kind of underground mind. Metaphors can inspire, but they also smuggle in assumptions: a network, we imagine, must be designed; coordination, we assume, requires a coordinator. Fungi make trouble for both assumptions. They build transport systems that are hard to kill, difficult to predict, and strikingly economical, all without a central controller. If that sounds like the sort of engineering slogan that ends with an exclamation point, it becomes more interesting when one asks a sharper question: what, precisely, are the rules by which a fungus lays down its roads?

An algorithm, in its plainest sense, is a procedure: a set of steps that turns local information into a global outcome. Most of the algorithms that run modern life are explicit and brittle. A logistics network is optimised by software whose designers choose what to value: speed, cost, reliability. A power grid is held together by control systems that monitor and adjust. A computer network routes packets according to protocols. When something goes wrong, engineers look for a fault, a bug, a mis-specified objective.

A fungus has no server room. What it has is growth: a restless exploration at the tips, and behind the tips a living plumbing system. Its "wires" are hyphae, thin tubes filled with fluid, partitioned by cross-walls in some groups and open in others. When hyphae bundle together into thicker strands called cords (in many forest fungi), they behave less like loose hairs and more like cables. Inside, the fluid does not simply drift; it can be pushed and pulled by pressure differences generated by growth, evaporation, and local uptake. A mycelium is not a static web. It is a deforming body that continuously rebuilds itself.

This is the sense in which a fungal network can be said to compute: not by representing a problem in symbols, but by reshaping its own structure until the structure fits the constraints. Luke Heaton and colleagues (2010) captured part of this in a compelling idea: growth itself can induce mass flows through the network. As tips extend, they require water and solutes; as regions mature, they thicken or regress. If fluid is effectively incompressible (as water largely is), then a local change in volume and demand can create pressure changes that propagate far away. Coordination emerges not because the fungus "knows" the whole network, but because the physics couples distant parts. One can tug a single strand in a spider web and feel the tension elsewhere; a fungus does something similar with fluid and pressure, except the web is alive and can reinforce the strands that carry the most traffic.

Seen from this angle, the underground looks less like an internet of messages and more like a transport authority under constant reconstruction. Paths that are used become wider; paths that are not used decay. The profound point is that such reinforcement can be achieved with local feedback: thicken where flow is high, prune where flow is low. The global consequence is a network that tends to concentrate capacity where it is most needed, while still keeping alternative routes. In human terms, the fungus "invests" in well-travelled corridors.

Network scientists did not arrive at this insight by staring at mushrooms. One of the most famous popular demonstrations of "natural algorithms" comes from an organism that is not a fungus at all, but a slime mould, *Physarum polycephalum.* It is a single-celled amoebozoan that forms sprawling, vein-like networks as it forages. In 2010, a team led by Toshiyuki Nakagaki staged an experiment that has since become an emblem for bio-inspired design. They placed oat flakes (food) on a plate in positions corresponding to cities around Tokyo and inoculated *Physarum* near the location of Tokyo itself. Over time, the organism grew a network of tubes connecting the food sources. The resulting pattern bore a remarkable resemblance to the Tokyo rail system: not identical, but comparable in its balance of efficiency and redundancy.

The viral lesson drawn from this was often simplified: slime mould "solves" the subway. The more honest lesson is subtler and more useful. The organism does not solve a map; it responds to local gradients of food and to its own internal flows. Tubes with greater flow become thicker; thinner tubes shrink. In a variable environment, this reinforcement-and-pruning rule yields networks that approximate efficient routes while maintaining alternative paths. A separate, more formal strand of work, including mathematical analysis by Vincenzo Bonifaci and colleagues (2011), showed that under certain models of flow-dependent reinforcement, networks can converge toward shortest paths in graphs. That matters because it turns a charming demonstration into a general principle: one can write down a local feedback rule and prove something about the global pattern it tends to produce. The proof does not claim the organism is aiming for a shortest path; it shows that "aiming" is not required for the outcome.

Fungi are different organisms with different constraints, but they share the key ingredients that make such dynamics plausible: flow through tubes, growth that responds to local conditions, and the ability to reallocate biomass. If you watch a fungus colonise a wood chip in a petri dish, it often begins as a diffuse fan of hyphae, then gradually differentiates: some strands thicken into cords, others fade. After a disturbance—a cut, a drought, a competitor—the network can reroute. That rerouting is not a decision in the human sense. It is a consequence of competing pressures: minimise the energetic cost of building and maintaining tubes, and yet keep the ability to reach new resources, survive damage, and avoid catastrophic isolation.

Engineering, too, lives inside competing pressures. A network that is perfectly efficient under normal conditions can be fragile; a network that is maximally resilient can be wasteful. The art is in negotiating the trade-off. What biology offers is not a single optimal shape but a family of solutions that shift with constraints. Cassidy Mentus and Marcus Roper (2020) explored this idea by modelling transport networks in which objectives include not only efficiency but also mixing—how well a network blends flows from different sources. Under strict constraints, the best networks can become tree-like, funneling flow along minimal paths. Under relaxed constraints or when mixing becomes valuable, loopy structures can be favoured. A loop is expensive in material but priceless when a link breaks; it also allows substances to circulate rather than simply travel outward and back. Anyone who has watched a city grid cope with road closures has an intuitive sense of why loops matter.

This is where fungal networks start to look like a rich design library. Many forest fungi build corded networks that resemble braided ropes with cross-links: not a neat lattice, but not a pure tree either. They often keep redundant connections, yet not so many that the system becomes a heavy, maintenance-draining carpet. Their "design choices" are written in hyphal thickness, branching angles, and the decision (triggered by local cues) to keep a cord alive or to let it wither.

A historical anecdote from the early days of computing makes the resonance clearer. In the 1960s, the US Department of Defense funded research that eventually led to ARPANET, the ancestor of the modern internet. One of its guiding ideas was resilience under damage: if part of the network is destroyed, information should still get through by rerouting. This required distributed routing rather than a single central switchboard. The irony is that the internet, which later gave us the "wood-wide web" metaphor, was itself designed by borrowing a principle that nature had already mastered: decentralised robustness. The fungus does not route packets, but it does face the same problem in a different currency. A cord is a commitment of biomass; if it is cut, the fungus must either abandon the branch or rebuild an alternative. Distributed reinforcement is one way to avoid putting all your resources into a single vulnerable trunk line.

Bio-inspired engineering often fails when it treats biology as an oracle rather than a tutor. The slime mould did not *invent* the Tokyo subway; humans built the subway using surveys, politics, budgets, and history. The point of the experiment is that under some conditions, very simple local rules can generate a network whose structure overlaps with a human-engineered optimum. That overlap can be mined for ideas: ways to balance cost and redundancy, ways to adapt when the environment changes, ways to build networks that "heal" without a central repair crew.

The same caution applies when borrowing from fungi. A fungal network is optimised for fungal fitness, not for human goals. It values survival, resource capture, and reproduction. It also runs on materials and timescales unlike ours: hyphae can be micrometres wide; cords thicken over hours to days; the penalty for failure may be starvation. Still, the underlying logic—local feedback producing global organisation—translates surprisingly well into mathematics, and mathematics translates well into code.

Consider the practical problem of designing supply chains under uncertainty. A company wants to ship goods from factories to warehouses to stores. Roads close, ports congest, demand spikes, fuel prices fluctuate. Traditional optimisation tries to calculate the best network given a model; when the model is wrong, the network can be brittle. A fungal-inspired approach would treat the network as plastic: continuously adjust capacities based on observed flows, reinforcing routes that carry reliable traffic and allowing underused routes to shrink, while keeping enough redundancy to avoid collapse when a single route fails. In software, this can take the form of adaptive routing algorithms and self-tuning capacities. In hardware infrastructure, it can motivate modular systems that can be reconfigured quickly.

Power grids face a related challenge. Electricity must be balanced in real time. A grid designed only for average conditions can fail dramatically under extreme events. Engineers increasingly explore "self-healing" grids: systems that detect faults and re-route power automatically. The conceptual rhyme with biological networks is close. When a fungal cord is severed, pressure gradients and local shortages can trigger regrowth or strengthen alternative paths. The fungal system senses failure through consequences, not through an abstract alarm. That difference matters: consequence-based adaptation can be robust when you cannot reliably predict every failure mode.

One reason fungi are particularly suggestive models is that they blur the line between network and organism. In human engineering, a network is often treated as an overlay on geography: roads on land, wires on poles, rails on sleepers. In fungi, the network is the body. Maintenance is metabolism. Expansion is growth. Repair is regrowth. These are not separate departments. That integration forces an honesty about costs. A cord that is thicker costs more to build and keep; if it does not pay its way in flow, it dies back. Engineers can learn from that discipline by designing networks where costs are not hidden, where unused redundancy is recognised as a liability unless it provides measurable resilience.

At the centre of bio-inspiration, the Tokyo experiment remains a touchstone because it can be shown, at a glance, as an overlay: a human network and a biological one converging on similar solutions. The most useful way to view it is not as a competition but as a conversation between constraints.

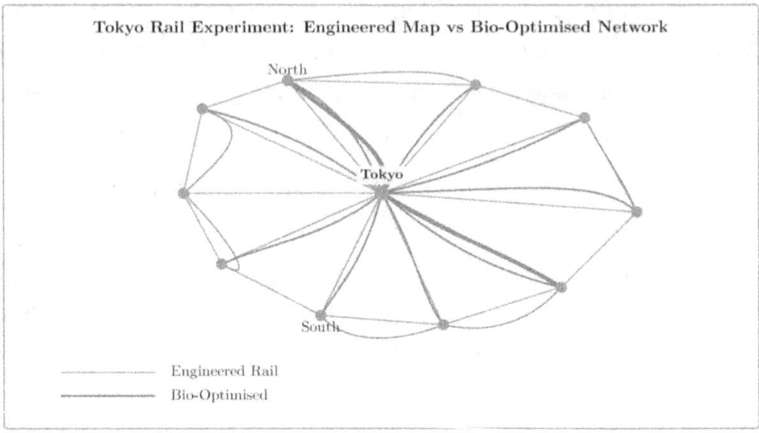

The figure is deliberately schematic. It matters less whether each node corresponds to a particular city and more that two different processes—human planning and biological reinforcement—can produce overlapping structural motifs: trunk routes that carry high flow, peripheral loops that provide redundancy, and shortcuts that appear when they pay for themselves.

A sceptic might object that this is merely a curiosity: a pretty overlay with little practical consequence. Yet the last two decades of network science suggest that the real consequence is conceptual. Biological networks force a rethinking of optimisation as something that can be distributed, incremental, and resilient to imperfect information. That is exactly the terrain where modern infrastructure struggles: climate-driven extremes, geopolitical shocks, cascading failures, and the limits of central prediction.

Fungi also teach restraint. The phrase "optimised network" is often used as if there were a single scalar target. In reality, optimisation is always plural. A fungal network that is excellent at rapid exploration may be poor at long-term transport; one that is excellent at conserving energy may be fragile to cuts; one that mixes contents well may be slower to deliver a concentrated resource to a specific sink. Engineers face the same plurality. The lesson is not "copy the fungus," but "write down your objectives, then ask what local rules might build and maintain a network under those objectives without constant top-down control."

There is a human comfort in believing that intelligence must look like deliberation. Fungal networks offer a different comfort: that good outcomes can arise from stubborn, local persistence and from feedback loops that do not require comprehension. A cord thickens where flow makes it worthwhile. A link disappears where it is not used. A detour becomes a main route when disruption forces traffic through it. That is not wisdom in the sentimental sense, but it is a kind of practical competence, earned by physics and tested by scarcity.

If someone insists on calling it "computation," it helps to be precise. The network is not solving Tokyo, or solving logistics, or solving the forest. It is solving itself: adjusting its own body to the demands imposed by resources, competitors, and the simple requirement that fluid and nutrients must get from where they are to where they are needed. In that self-adjustment lies a design principle with unusual honesty: the map is not separate from the territory, because the territory is inside the organism.

10.3 Stewards of the Invisible

A forester can walk a stand with a diameter tape and come away with a story: growth rates, merchantable volume, the next thinning, the next harvest. The story is not wrong; it is simply written in the most visible ink. "Network science asks what happens when the forest is measured not only as trunks and crowns, but as a set of relationships in space: who is connected to whom, where the bottlenecks are, how quickly a disturbance cascades, how long recovery takes, and which parts of the system act like slow, stabilising memory."

Once that perspective is taken seriously, it becomes difficult to treat soil as mere substrate. Soil is an ecosystem with architecture. Some of that architecture is mineral and physical: pores, aggregates, root channels, the capillary pathways that hold water against gravity. Some is biological: bacterial biofilms, the grazing loops of protozoa, the microscopic churn of nematodes. Fungal networks sit in the middle, not as benevolent puppeteers, but as living infrastructure that links carbon aboveground to nutrients belowground, and that can persist (or fail) on timescales that matter for climate and forestry alike.

The simplest stewardship lesson is also the least glamorous: *don't confuse a compelling narrative with an actionable target.* Karst and colleagues (2023) argued that knowledge about common mycorrhizal networks in forests remains too unsettled to justify management interventions aimed specifically at "enhancing CMNs," especially if the goal is framed in terms of parental altruism or guaranteed seedling rescue. That caution is not a call to ignore fungi; it is a call to be precise about what is being managed and how it is being measured.

Precision begins with vocabulary. "Network" can mean several different things, and the differences matter because methods that can detect one kind of network may be blind to another. A single word can hide a chain of assumptions: that hyphae are continuous from one plant to another; that the same fungal individual is involved; that the pathway is responsible for a measured transfer; that the transfer changes growth. When those assumptions are bundled together, the science becomes fragile: one weak link can make the entire story collapse.

A small amount of definitional clarity can do more for stewardship than a shelf of metaphors. Ecologists have increasingly distinguished between the existence of a shared fungal community and the existence of a direct hyphal connection that can serve as a transport route. The details vary across authors, but the practical message is consistent: match the claim to the evidence.

Term	*Practical meaning for stewardship and measurement*
CMN	A common mycorrhizal network: hyphal connections linking multiple plants through mycorrhizal fungi. Strong claims about direct resource transfer usually require evidence that the same fungal individual connects the plants.
CMN-HC	A stricter, hyphal-continuity emphasis: the claim is specifically that an intact hyphal pathway provides the route (not merely a shared fungal presence in the same soil). This framing pushes researchers to test continuity and to rule out alternative pathways.
CFN	A broader "common fungal network" idea: plants may be linked indirectly through a shared fungal community and its effects on soil chemistry, pathogens, and nutrient cycling, even when direct hyphal continuity is not established. This is often more relevant to whole-stand function than to pinpoint transfers between individuals.

With those distinctions in hand, stewardship stops being a hunt for a single magical property ("the wood-wide web") and becomes a set of grounded questions. What fungal lineages are present? How diverse are they? How intact is the soil structure that allows hyphae to persist? How frequently is the system reset by disturbance? Which interventions help, and under what conditions?

A useful historical anecdote comes from the nineteenth century, before "networks" became a fashionable word. In 1885, the German botanist Albert Bernhard Frank introduced the term *mycorrhiza* to describe the intimate association between roots and fungi. At the time, this was not a romantic discovery but a practical puzzle: why did certain trees grow poorly when transplanted, even in seemingly adequate soil? The early answer was not "the forest shares"; it was "the plant's underground partnerships have been broken." Modern forestry has replayed that puzzle repeatedly, often without realising it. When soil is scraped, compacted, or sterilised; when planting stock is raised in conditions that do not resemble the target site; when a stand is replanted as a uniform monoculture; the question is not only whether trees can establish, but whether the invisible partnerships that make establishment efficient are present and functional.

The climate relevance is immediate. Trees draw down carbon as they grow, but much of the long-term carbon story is written in soil. Fungal tissues contribute to soil organic matter; fungal necromass (dead fungal material) can persist, becoming part of stable carbon pools, especially when it binds to minerals. Mycorrhizal fungi also influence how nitrogen and phosphorus move through ecosystems, which in turn shapes plant growth and the balance between carbon captured and carbon respired back to the atmosphere. In a warming world, the stability of these belowground processes matters as much as the pace of planting.

That does not mean every fungal-friendly action is automatically climate-smart. It means that interventions need to be judged with the same sober eye one would apply to aboveground management: define an objective, specify a mechanism, measure outcomes, and accept that context will often dominate.

Consider mycorrhizal inoculation, the deliberate addition of fungi to soils or seedlings. It is tempting to talk about inoculation the way one talks about vaccination: a beneficial microbe added to confer resilience. But the analogy breaks down quickly. Vaccines target a specific pathogen and a specific immune response. Mycorrhizal inoculation introduces living organisms into a community already full of living organisms, and the introduced fungi must compete, persist, and form effective partnerships under local conditions.

Meta-analyses of field studies suggest that inoculation can, on average, increase plant biomass, but the effect is variable and depends on factors such as plant functional type and soil nutrient status. In practice, that variability is not an inconvenience; it is the main message. In nutrient-poor soils, a well-matched mycorrhizal partner can act like an extension of the root system. In nutrient-rich soils, the plant may have little incentive to pay carbon for nutrients it can acquire cheaply, and the partnership can slide along the mutualism—parasitism continuum. Inoculation can also fail simply because the local fungal community outcompetes the introduced strains, or because the introduced strains are not adapted to the site's pH, moisture regime, or seasonal rhythms.

The stewardship implication is not "never inoculate." It is "treat inoculation as an ecological intervention, not as a product." An intervention deserves a diagnosis. If a site is severely disturbed, compacted, eroded, or chemically altered, fungal limitation may genuinely constrain establishment, and inoculation may help. If a site retains intact soil and nearby fungal sources, disturbing it further to apply inoculum may be counterproductive. There is no shame in that uncertainty. There is only risk in pretending it does not exist.

The same realism applies to old growth. Old forests are often defended on moral grounds, as if age itself conferred virtue. A network view provides a sturdier defence: old growth contains accumulated structure, both above and below ground, that cannot be rebuilt on human timescales by simply planting seedlings. Large trees buffer microclimate by shading the soil and reducing temperature extremes. Deep, undisturbed soils maintain pore networks that store water and allow roots and hyphae to explore. Dead wood provides habitat and slow-release nutrients. Fungal communities in old forests can be diverse and spatially complex, reflecting centuries of continuity and disturbance history.

Continuity is a form of wealth. It is also easy to spend and hard to replace.

One way to make this more concrete is to imagine a clear-cut not as the removal of trees but as the sudden loss of a transport-and-exchange infrastructure. Aboveground, the canopy is gone. Belowground, the carbon supply to fungi collapses in patches; hyphal networks fragment; soil temperature and moisture swing more dramatically; erosion risk rises; the microbial community shifts. Some fungal species persist, some rebound quickly, and some may be lost locally for a long time. Reforestation can restore tree cover, but rebuilding a mature belowground community is not like rebuilding a fence. It is closer to rebuilding a city's informal economies, its trust networks, its accumulated skills: possible, but slow, and not guaranteed to follow the same trajectory.

If stewardship begins locally, it now scales globally. In recent years, projects such as the Society for the Protection of Underground Networks (SPUN) have tried to map mycorrhizal fungal diversity and model its distribution at planetary scale, producing what is sometimes called an "Underground Atlas." The ambition is not to romanticise fungi, but to make them legible to decision-makers who are used to maps: layers of predicted richness, uncertainty estimates, and overlaps with land-use and protected areas.

The promise of such mapping is practical. Conservation often follows what can be counted and shown. Birds and mammals have long benefited from this bias because they are visible and charismatic. Fungi are neither. When fungi do enter public conversation, they often arrive through a single story—the wood-wide web, the mother tree—rather than through the quieter, less cinematic truth that fungi are a vast reservoir of functional diversity that underpins nutrient cycles, soil structure, and plant resilience. A map shifts the conversation from anecdote to pattern. It asks uncomfortable questions. Are regions predicted to harbour high mycorrhizal diversity being converted to agriculture or development? Do protected areas overlap the places where belowground diversity is concentrated? Are restoration projects being planned in ways that preserve local soil communities, or do they treat soil as interchangeable?

Such questions can feel abstract until they collide with an event. Imagine a drought year in a managed forest: seedlings planted after harvest struggle, not only because rain is scarce, but because soil has compacted and the fine-scale pathways for water and hyphal growth are reduced. Or imagine an extreme rainfall event: erosion strips topsoil from slopes, carrying away the very layer richest in organic matter and microbial life. In both cases, the visible damage is aboveground, but the capacity to recover is largely belowground. A forest can be replanted in a season. A soil community may take decades to reassemble, and some losses may be effectively permanent on management timescales.

Stewardship, in this light, becomes less about heroic interventions and more about preserving options. Options live in diversity. A diverse fungal community is not a guarantee of any single outcome; it is a reservoir of strategies. Some fungi are better at scavenging phosphorus, others at accessing organic nitrogen, others at tolerating drought, others at surviving fire or cold. In a stable climate, a narrower set of partnerships might be sufficient. In an unstable climate, betting everything on a single strategy is an invitation to failure.

The phrase "climate resilience" is often used as if resilience were a trait, like height or wood density. Resilience is better understood as a property of a system's responses: how it absorbs shocks, how it reorganises, and whether it retains function. Network science offers a useful way to see why diversity helps. A system with many nodes and multiple pathways can reroute when a connection breaks. A system with uniformity and few alternatives can suffer cascading collapse. Underfoot, fungal diversity supplies alternative pathways of nutrient acquisition and alternative patterns of association with plants. Aboveground, mixed-species stands often spread risk across drought tolerance, rooting depth, and phenology. The two layers interact. Stewardship that focuses only on the visible layer is like reinforcing a bridge deck while ignoring corrosion in the supports.

None of this requires believing that forests are altruistic. The underground is not a moral community. It is a living infrastructure assembled from countless transactions, conflicts, and compromises, shaped by local feedbacks and physical constraints. That infrastructure can still be cared for. Care, in the human sense, is not something we expect from fungi. It is something we bring to our decisions when we recognise what is easy to destroy and hard to replace.

A handful of soil crumbled between fingers rarely looks like an archive. Yet it is one: of past climates, past fires, past roots, and the slow work of organisms that never appear in a timber inventory. To be a steward of the invisible is to treat that archive as part of the forest itself, and to manage with the patience that an underground timescale demands.

www.ingramcontent.com/pod-product-compliance
Lightning Source LLC
LaVergne TN
LVHW022200200926
848481LV00006B/1791
* 9 7 9 8 8 9 6 6 5 9 9 0 7 *